WATER RESOURCE PLANNING, DEVELOPMENT AND MANAGEMENT

U.S. INTERNATIONAL WATER AND SANITATION ASSISTANCE

STRATEGY AND ASSESSMENT

Water Resource Planning, Development and Management

Additional books in this series can be found on Nova's website under the Series tab.

Additional e-books in this series can be found on Nova's website under the eBooks tab.

WATER RESOURCE PLANNING, DEVELOPMENT AND MANAGEMENT

U.S. INTERNATIONAL WATER AND SANITATION ASSISTANCE

STRATEGY AND ASSESSMENT

CONSTANCE ALVARADO
EDITOR

nova publishers
New York

NOTICE TO THE READER

The Publisher has taken reasonable care in the preparation of this book, but makes no expressed or implied warranty of any kind and assumes no responsibility for any errors or omissions. No liability is assumed for incidental or consequential damages in connection with or arising out of information contained in this book. The Publisher shall not be liable for any special, consequential, or exemplary damages resulting, in whole or in part, from the readers' use of, or reliance upon, this material. Any parts of this book based on government reports are so indicated and copyright is claimed for those parts to the extent applicable to compilations of such works.

Independent verification should be sought for any data, advice or recommendations contained in this book. In addition, no responsibility is assumed by the publisher for any injury and/or damage to persons or property arising from any methods, products, instructions, ideas or otherwise contained in this publication.

This publication is designed to provide accurate and authoritative information with regard to the subject matter covered herein. It is sold with the clear understanding that the Publisher is not engaged in rendering legal or any other professional services. If legal or any other expert assistance is required, the services of a competent person should be sought. FROM A DECLARATION OF PARTICIPANTS JOINTLY ADOPTED BY A COMMITTEE OF THE AMERICAN BAR ASSOCIATION AND A COMMITTEE OF PUBLISHERS.

Additional color graphics may be available in the e-book version of this book.

Library of Congress Cataloging-in-Publication Data

ISBN: 978-1-63485-066-7

Published by Nova Science Publishers, Inc. † New York

CONTENTS

PREFACE

Water is a foundational element of development and by its nature, a basic and essential resource. A great challenge confronts certain regions, nations and individuals in the world in the form of having enough sustainable water. As discussed in this book, the US Agency for International Development's (USAID) strategy presents a guide to investments in water programming and sanitation efforts, informs country development strategies, guides decision making on budgeting and resource allocation, and highlights priority regions and areas for water programs. The book describes recent activities and funding, assesses USAID missions' efforts to plan and implement activities, and assesses USAID's monitoring of activities.

In: U.S. International Water ...
Editor: Constance Alvarado

ISBN: 978-1-63485-066-7

Chapter 1

USAID WATER AND DEVELOPMENT STRATEGY 2013-2018*

U.S. Agency for International Development

LIST OF ACRONYMS

BFS	Bureau for Food Security
BRM	Office of Budget and Resource Management
CATS	Community Approaches to Total Sanitation
CDCS	Country Development Cooperation Strategy
CGIAR	Consultative Group on International Agricultural Research
CLTS	Community-Led Total Sanitation
CMM	Conflict Management and Mitigation
DA	Development Assistance
DCA	Development Credit Authority
DCHA	Bureau for Democracy, Conflict and Humanitarian Assistance
DIV	Development Innovation Ventures
DRR	Disaster Risk Reduction
E3	Bureau for Economic Growth, Education and Environment
E3/W	Office of Water in E3 Bureau
ESF	Economic Support Funds

* This is an edited, reformatted and augmented version of a document issued by USAID, May 2013.

ESP	Environmental Services Program
EXIM	Export-Import Bank
FTF	Feed the Future
FY	Fiscal Year
GH	Bureau for Global Health
HIP	Hygiene Improvement Project
IDA	International Disaster Assistance
IR	Intermediate Results
IWRM	Integrated Water Resource Management
JMP	Joint Monitoring Program
MDG	Millennium Development Goal
NEAT	Nepal Economic Agriculture and Trade
NGO	Non-Governmental Organization
NOAA	National Oceanic and Atmosphere Administration
OFDA	Office of U.S. Foreign Disaster Assistance
OPIC	Overseas Private Investment Corporation
PPD-6	Presidential Policy Directive on Global Development
PPL	Bureau for Policy, Planning, and Learning
PWRF	Philippines Water Revolving Fund
QDDR	Quadrennial Diplomacy and Development Review
RANET	Radio and Internet Technologies for the Communication of Weather and Climate Information for Rural Development
SCADA	Supervisory Control and Data Acquisition
SEEDS	School Environment and Education Development for Somalia
SO	Strategic Objective
SPLASH	Schools Promoting Learning Achievement through Sanitation and Hygiene
SWSS	Sustainable Water Supply and Sanitation
WASH	Water Supply, Sanitation and Hygiene
WATER	Water, Sanitation, and Hygiene Transformations for Enhanced Resilience

EXECUTIVE SUMMARY

This is the US Agency for International Development's (USAID) first global Water and Development Strategy. It is intended to provide a clear understanding of USAID's approach to water programming. This Strategy emphasizes how sustainable use of water is critical to save lives, promote sustainable development, and achieve humanitarian goals.

Projections are that by 2025, two-thirds of the world's population could be living in severe water stress conditions. This stress adversely affects individuals, communities, economies, and ecosystems around the world, especially in developing countries. Ensuring the availability of safe water to sustain natural systems and human life is integral to the success of the development objectives, foreign policy goals, and national security interests of the United States.

To address global water-related development needs, this Strategy provides an increased focus for USAID's water programs that is sustainable, works through host country systems, uses emerging science and technology, and learns from past efforts.

The goal of this Strategy is: To save lives and advance development through improvements in water supply, sanitation, and hygiene (WASH) programs, and through sound management and use of water for food security. To achieve this goal, the Strategy sets two strategic objectives (SOs):

SO1) Improve health outcomes through the provision of sustainable WASH. This will be achieved through a continued focus on providing safe water, an increased emphasis on sanitation, and support for programs that can be brought to scale and be sustained. Based on previously requested funding levels, USAID projects providing a minimum of 10 million persons with sustainable access to improved water supply and 6 million persons with sustainable access to improved sanitation over the next five years.

SO2) Manage water for agriculture sustainably and more productively to enhance food security. This will be achieved through increased emphasis on more efficient use of rainfall and improved efficiency and management of existing irrigation systems including private and farmer-owned micro-irrigation systems. The Strategy recognizes that the greatest and most cost-effective potential for crop yield increases are in rainfed areas. USAID will focus on increasing irrigated agriculture in select countries, including expanding irrigation in a responsible, sustainable, and climate-resilient way.

Historically, USAID budget allocations for water programs have been made in four thematic areas: (1) WASH; (2) Water Resources Management; (3) Water Productivity; and (4) Disaster Risk Reduction. Between Fiscal Years 2003-2011, USAID annually allocated an average of $452M to water activities. Over this time period, USAID annually programmed an average of $318M to WASH, $61M to Water Resources Management, $57M to Water Productivty, and $16M to Disaster Risk Reduction.

Beginning in Fiscal Year (FY) 2014, the Strategy calls for all new USAID water programs to fully align with the two SOs of the USAID Water and Development Strategy.

Strategic Approach

To achieve these objectives, this Strategy:

- Furthers the strategic approach to water and development put forward by the Secretary of State in 2010 by emphasizing five 'streams of action' to address water issues.
- Advances activities consistent with the Senator Paul Simon Water for the Poor Act of 2005 including establishing criteria to designate high priority countries for increased investments to support access to safe water and sanitation.
- Addresses the impact of water problems for countries important to U.S. national security interests.
- Builds on the joint USAID and U.S. Department of State's 2008 Framework for Action, which called for improving access to water supply and sanitation, promoting better hygiene, improving water resources management, and improving water productivity in agriculture.
- Draws on USAID Forward and the USAID Policy Framework 2011-2015 by supporting host country systems, emphasizing an integrated approach to development, focusing where resources will be invested, programming resources selectively to ensure meaningful impact, leveraging science, technology, and innovation, promoting gender equality and female empowerment, and increasing partnerships.
- Gives priority to supporting the Presidential Initiatives of Global Health and Feed the Future, and supports other development objectives such as resilience, natural resource and coastal zone management, climate change efforts, humanitarian assistance, increasing access to education, and gender equality efforts.

- Will seek investments in longer-term monitoring and evaluation of its water activities in order to assess sustainability beyond the typical USAID Program Cycle and to enable reasonable support to issues that arise subsequent to post-completion of project implementation.
- Emphasizes an integrated water resources management approach to development.
- Increases attention to sanitation in WASH programs and encourages multiple use systems of water for agriculture.
- Builds on USAID's comparative advantage in capacity building and governance by emphasizing policy reform, strengthened enabling environments and institutions, participatory governance, and innovative financing.
- Links humanitarian and development efforts more effectively by supporting programming of WASH funds in emergency situations.
- Applies the USAID Gender Equality and Female Empowerment Policy by supporting gender sensitive approaches to empower women in the development and implementation of water-related programs.

USAID intends to address the global challenges of water in close cooperation with non-governmental and civil society organizations that undertake the critical front-line responsibility of developing and implementing water programs. This includes working with advocacy groups that bring both knowledge and passion to the challenge, with governments that are dedicated to providing a better life for their citizens, with communities that best understand the challenges and solutions, with universities that are creating innovative solutions today, with the private sector that can build a new global economy while supporting sustainable development, and with international development and financial institutions which provide essential program development implementation and financial support.

USAID seeks to be both a leader and a partner in the global effort to elevate the importance of water across all development objectives.

1. Introduction

Achieving water security for regions, nations, and individuals is one of the greatest development challenges confronting the world today. By its nature, as a basic and essential resource, water considerations cut across nearly every aspect of USAID programming.

The purpose of the Strategy is to guide USAID investments in water programming, to inform the development of Country Development Cooperation Strategies (CDCS), to guide decision-making on budgeting and resource allocation, and to highlight priority regions and program areas for water programs. Over the duration of the Strategy there will be greater focus and selectivity among USAID water programs leading to greater development impact.

Strategic Focus

USAID has three overarching development objectives that address the Presidential Initiatives of Climate Change, Food Security, and Global Health. This Strategy focuses on two key development objectives related to effective and multiple uses of water resources: water for health and water for food. USAID addresses the climate change and water linkage in its Climate Change and Development Strategy (2012-2016).[1] This Strategy responds to the need for USAID to focus investments and identify priorities within the wider role that water and watershed management play toward energy, conflict, climate change, education, bio-diversity, ecosystems, and economic growth. This Strategy specifically endorses the principles and proven approaches of integrated water resources management (IWRM) and encourages the use of all appropriate technologies and tools in achieving those objectives.

Key Development Challenges

Population growth, increased demand for and rising cost of energy, increased urbanization, watershed and environmental degradation, natural disasters, conflict, climate change, and weak water governance are putting water resources under increasing pressure. Projections are that by 2025, two-thirds of the world's population could be living in severe water stress conditions.[2]

This stress adversely affects individuals, communities, economies, and ecosystems around the world, especially in developing countries. It also underscores why it is so critical to properly manage the scarce freshwater resources upon which human life depends.

Inadequate Access to Safe Drinking Water and Sanitation

Millennium Development Goal (MDG) 7C calls to halve by 2015 the proportion of the population without sustainable access to safe drinking water and basic sanitation. While the safe drinking water target was met in 2010, 783 million people still do not have access to safe drinking water, and major issues related to equity of access, water quality, and sustainability of water supplies remain.[3] In addition, the world is not on track to meet the sanitation target as approximately 2.5 billion people still lack access to improved sanitation.[4]The challenges and solutions vary significantly by region; for example, sub-Saharan Africa has the highest proportion of people without sufficient sanitation facilities, while South Asia has the largest number of people practicing open defecation.

Often the burden of inadequate access to water and sanitation falls heavily on women and girls. Examples of this are evident throughout the developing world. Two concerns of particular importance are reducing the many hours women and girls spend seeking water for their families which often put their safety at risk and addressing the different sanitation needs of women and adolescent girls which have direct impacts on maternal mortality and morbidity.

Lack of access to safe water and sanitation services has direct health implications as nearly two million people – the vast majority of whom are children under five – die from diarrhea each year. Nearly 88 percent of diarrhea is attributed to unsafe drinking water, inadequate sanitation, and poor hygiene, and is preventable by known interventions.[5]

The Importance of Meeting Sanitation Goals

Conventional sanitation, a flush toilet connected to a centralized sewer system, is available for only a small fraction of people in developing nations. Forty percent of the world's population use unsafe toilets or practice open defecation. The consequences of unsafe sanitation can be devastating and last a lifetime. Every year, food and water tainted with fecal matter cause up to 2.5 billion cases of diarrhea among children under five, resulting in 1.5 million child deaths. Chronic diarrhea can hinder child development by impeding the uptake of essential nutrients that are critical to the development of children's minds, bodies, and immune systems. Studies show that hand washing, improved sanitation, and improvements in household water quality significantly reduce the risk of diarrhea.[6]

The impact that clean and safe sanitation can deliver is transformational. The economic benefits of improved sanitation can increase productivity, reduce healthcare costs, and prevent illness, disability, and early death. People who have access to clean, safe, and convenient sanitation services also experience greater dignity, privacy, and security. This is especially important for women and girls, who may miss work or school when they are menstruating or risk sexual assault when they do not have access to safe sanitation facilities. This importance extends not just to the existence of sanitation facilities, but also to the physical location and design in a school/workplace that has an impact on safety, security, and use.

Limited Access to Fresh Water for Food Security

The interdependence between food security and sustainable water resource management is significant. Globally, agriculture consumes 70 percent of available freshwater resources and is often used in irrigated systems that are inefficient and environmentally unsustainable.[7] Global population growth projections of two to three billion people over the next 40 years, combined with changing diets, are expected to increase food demand 70 percent by 2050.[8] However, as the largest user of water, food production also represents the largest unknown factor of future water use in terms of future global water demand. Significant efficiency gains are possible. Based on current use patterns, agricultural water consumption will increase by approximately 19 percent to feed a larger and richer global population of 9.1 billion people.[9]

The Impact of Climate Change on Water Resources

The impacts of climate change can be seen from oceans to mountain tops. Rapid glacier melt and decreased snowpack will increase variability of stream flows and may contribute to a long term decline in supplies. Rising sea levels will exacerbate saltwater intrusion into many river deltas, salinizing productive/cultivatable land and impacting drinking water. Stronger storm surges could flood large tracts of coastal areas, where a significant number of people and critical infrastructure are located. The effects of climate change in combination with other drivers of stress on water resources will increasingly restrict access to safe water and sanitation and make food security even harder to achieve.

The risks of climate change to rainfed agriculture could be particularly acute. Increased rainfall variability and reduced natural storage will likely reduce crop yields by producing higher rates of run-off, escalate soil desiccation at critical times, and shift rainfall patterns and the timing of rainy seasons. Resilience to climate change and other pressures is an emergent property of a functioning ecosystem, and water programs need to consider climate impacts. This Strategy seeks to strengthen adaptation and resilience to climate change, in accordance with the USAID Climate Change and Development Strategy, through increased efforts to support integrated water and watershed management practices. The USAID resiliency policy will further guide Agency programming and development approach.[10]

The Significant Energy Requirements for Water

Every drop of water that has to be pumped, moved, or treated to meet health and food needs requires energy. Inefficiency and waste in the use of water for industrial production increases energy demands and raises emissions. Effective co-management of water and energy, including the integration of water, food and energy programs, as well as support for and development of technology, can lead to significant returns on investment.

Water as a Potential Source of Conflict

As indicated by the 2012 Intelligence Community Assessment on Global Water Security, water problems will contribute to instability in countries important to U.S. security interests.[11] Water security is an increasingly important component of the U.S. Government's diplomatic and development efforts to promote peace and security within and between key countries and around trans-boundary river basins. Growing demands on limited fresh water, degradation of fresh water quality, and greater variability in rainfall patterns are potential drivers of tension. Competition and disputes over water and watersheds exist in many places around the world. The causes and nature of these disputes vary widely; from small-scale clashes over pasture and water, to urban protests over changes in water pricing schemes, to sub-regional disputes between provinces over water for agriculture or hydropower, to upstream/downstream countries competing for a share of an increasingly limited water supply.

As water becomes scarcer at community water points, women and children who gather the water may find themselves at the forefront of inter- and intra-community based conflict as they compete against each other for access to scarce water resources. The U.S. National Action Plan on Women, Peace and Security, for instance, calls for building resilience through assistance that supports women's roles in the management of natural resources, mitigation of resource related conflict, and adaptation to climate change in fragile and conflict-affected states.

2. Strategy Context

USAID supports integrated water resource management as a vital approach to achieve global development objectives. USAID's engagement in the water sector will continue to reflect the Senator Paul Simon Water for the Poor Act of 2005, annual appropriations directives, the Presidential Policy Directive on Global Development (PPD-6), the Quadrennial Diplomacy and Development Review (QDDR), USAID Forward, and the USAID Policy Framework 2011-2015, as well as global targets such as the Millennium Development Goals (MDG). The Strategy raises the importance of water programming across the Presidential Initiatives of Global Health, Feed the Future, and Global Climate Change.[14]

With limited resources in the face of significant challenges, the Strategy responds to PPD-6's call for selectivity and focus by prioritizing the water needs for health and food security. USAID's resources are most effective by leveraging resources at the country level, and by using emerging science and technology to promote innovation. As such, the Strategy emphasizes sustainable solutions that work through host country systems.

The Strategy applies selectivity by focusing resources in fewer countries, sub-national regions, sectors or programs so that USAID can have the greatest possible development impact. In addition, the Strategy encourages shifts in program emphasis to include increasing attention to sanitation in WASH programs, supporting multiple use systems of water supply for agriculture and, when appropriate, integrating water programming with multiple development objectives.

The Strategy builds on the joint USAID and U.S. Department of State's 2008 Framework for Action, which called for improving access to water supply and sanitation, promoting better hygiene, improving water resources management, and improving water productivity in agriculture.[15] The Strategy furthers the strategic approach to water and development put forward by the Secretary of State in 2010 by emphasizing five' streams of action' to address water issues. These streams include building capacity at the local, national, and regional levels, elevating our diplomatic efforts, mobilizing financial support, harnessing the power of science and technology, and broadening the scope of our partnerships to leverage our efforts against the work of others.[16]

The Strategy also draws upon the extensive experience of public and private sector partners that are engaged in the financing, development, and support of USAID programs and development objectives. Among others, this includes non-governmental organizations (NGOs), businesses, faith-based organizations, universities and international organizations.

Implementation of the Senator Paul Simon Water for the Poor Act of 2005

One of the primary sources of guidance for water programming is the Water for the Poor Act. As amended to the Foreign Assistance Act, this legislation defines access to safe water and sanitation for developing countries as a specific policy objective of U.S. foreign assistance programs. Each year USAID, with the U.S. Department of State, reports to the U.S. Congress on progress achieved towards the goals of this Act. This Strategy supports this effort by advancing many activities consistent with the goals of the Act including:

- Developing quantitative metrics and indicators to measure results;
- Mobilizing and leveraging the financial and technical capacity of public and private entities through partnerships;
- Encouraging capacity building to strengthen the ability of host countries to develop, manage and implement water programs and practice watershed management;
- Supporting governance structures, regulations and policies to expand access to safe water and sanitation services;
- Protecting the supply and availability of safe drinking water;
- Establishing criteria to designate high priority countries for increased investments to support access to safe water and sanitation; and
- Supporting sound environmental management including the protection of watersheds.

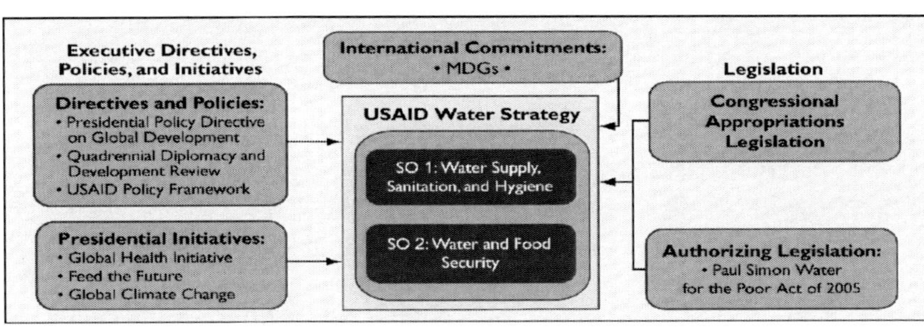

Support for Feed the Future

Feed the Future (FTF), the U.S. Government's global hunger and food security initiative, supports country-driven approaches to address the root causes of hunger and poverty and forge long-term solutions to food insecurity. Through FTF, the U.S. Government renewed its commitment to agriculture and economic growth and focused on harnessing the power of the private sector and research to transform agricultural development. Integrated water resource management is a critical component of the long term objectives of FTF to meet the challenges of global hunger.[17]

USAID Program Cycle

The USAID Program Cycle integrates Agency-wide polices and strategies, strategic planning, project design and implementation, and evaluation and monitoring. The Water Strategy represents one component of the Program Cycle, the setting of policies and strategies, within a broader cycle of an integrated approach to development. The Strategy provides Missions with strategic direction for the CDCS process and encourages Missions to formulate development objectives and multi-sector approaches to integrate water for health and water for food programming.

Historically, USAID budget allocations for water programs have been made in four thematic areas: (1) WASH; (2) Water Resources Management; (3) Water Productivity; and (4) Disaster Risk Reduction, Between Fiscal Years 2003-2011, USAID annually allocated an average of $452M to water activities. Over this time period, USAID annually programmed an average of $318M to WASH, $61M to Water Resources Management, $57M to Water Productivty, and $16M to Disaster Risk Reduction.

Consistent with the USAID Policy Framework, when designing water programs Missions are required to apply selectivity in regions and program areas, and to focus the volume of resources to have a meaningful, measurable, and lasting impact.

Based on the analysis and results framework of the CDCS, the Strategy will be operationalized through project design. The Project Design guidance outlines a series of steps each Mission should take when operationalizing the Strategy. Projects, defined as a set of interventions intended to achieve a discrete development result over a specific timeframe, in the water sector should align to the priorities outlined in the Strategy. To help determine whether projects are having the intended impact Missions should monitor the achievements of programs and projects, analyze results and track progress, use data to inform decision making, and communicate this information to stakeholders. Project design should utilize gender, youth, and community analyses to ensure that design, implementation, and measurement of impact specifically include women and children. All Program Cycle components applied to water programming contribute to improving the clarity, analytic rigor, and evidence with which requested resources connect to expected results. Programing contribute to improving the clarity, analytic rigor, and evidence with which requested resources connect to expected results.

USAID Budget Allocations for the Water Sector by Thematic Area Fiscal Years 2003 20 1 1 (Millions of Dollars)

USAID Budget Allocations for the Water Sector by Theme ($ Millions)	2003	2004	2005	2006	2007	2008	2009	2010	2011	Ave. 2003-2011
Drinking Water Supply, Sanitation & Hygiene (WSSH)	159.80	239.80	2 1 6.93	265.00	213.22	389.92	493.0 1	520.4 1	360.05	318
Water Resources Management (WRM)	105.70	82.50	60.73	56.00	27.41	58.58	41.24	47.20	67. 1 9	6 1
Water Productivity (WP)	115.60	68.40	45.35	22.50	17.39	38.91	45.30	53.11	109.30	57
Disaster Risk Reduction	20.60	1 0.00	6.76	5.84	5.65	2.20	50.55	2 1 .45	21.82	1 6
Grand Totals	401.70	400.70	329.77	349.34	263.67	489.61	630.11	642.17	558.36	452

3. Strategic Objectives 2013-2018

The overarching goal of this Strategy is: To save lives and advance development through improvements in WASH programs, and through sound management and use of water for food security. To achieve this goal USAID will pursue two SOs:

- SO 1 Improve health outcomes through the provision of sustainable WASH
- SO 2 Manage water in agriculture sustainably and more productively to enhance food security

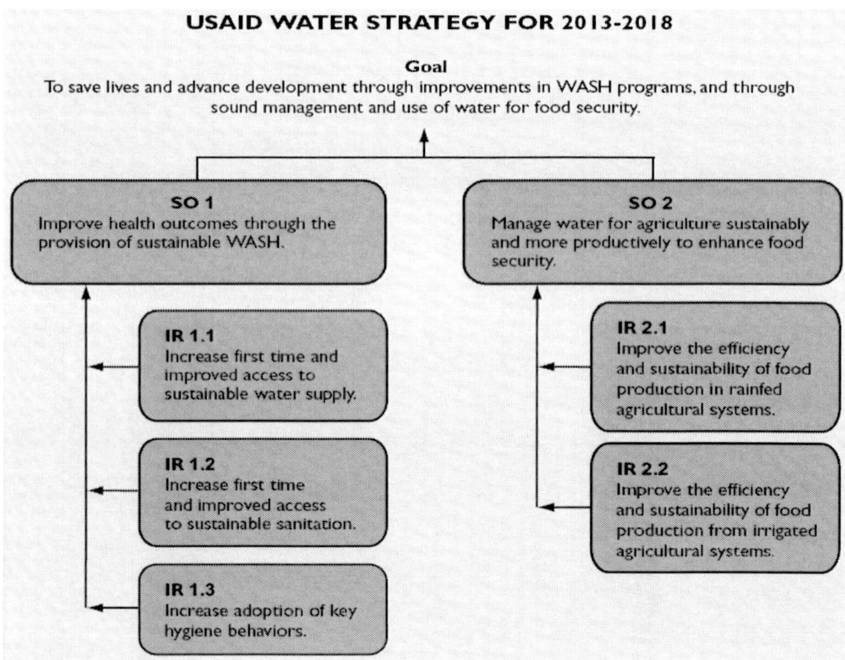

To accomplish the SOs, intermediate results (IRs) have been identified that will guide the planning and implementation of this effort.

SO 1 – Improve Health Outcomes through the Provision of Sustainable Wash

SO 1 Measures of Success 2013-2018

- Projects providing a minimum of 10 million persons with sustainable access to improved water supply and 6 million persons with sustainable access to improved sanitation.
- Key hygiene behaviors adopted in priority WASH countries

WASH investments can improve health and save lives, especially by reducing incidences of diarrheal disease in children under five years. By improving health outcomes through improved WASH practices, girls can have better opportunities for education and women can be less burdened in the home. WASH programs have significant economic and social benefits including reducing the time, labor and risk of fetching water, and reducing violence against women who use public facilities or defecate in the open after nightfall. Predictable access to water expands job options for both men and women and supports broad-based economic development. When safe water and sanitation are reliable and accessible, food security and nutrition can be improved. When wastewater is adequately treated, environmental quality, food safety, and human health are better protected.

Water, sanitation and hygiene approaches are most effective in reducing diarrhea and other water-related infections when implemented as a package of interventions. A balanced WASH program has three interdependent pillars: (1) hardware (e.g., water and sanitation infrastructure); (2) the promotion of behavior change; and (3) support to an enabling policy and institutional environment. USAID support will generally include interventions within all three pillars of the framework, with different levels of emphasis in each area as determined by the development context.

WASH Needs and Opportunity Criteria

WASH country needs criteria:

- The number of people and the percent of the population without access to improved sources of safe water and sanitation.
- The mortality rate and number of deaths of children under five years due to diarrhea.

WASH country opportunity criteria:

- The extent of host government commitment to WASH.
- The presence of supportive enabling environments.
- The opportunities for leveraging existing donor and/or private sector investments in the water sector.

Country Selectivity — Needs and Opportunity Criteria

USAID's WASH activities in FY 2012 were spread across 62 bilateral programs, regional platforms, and centrally funded programs. USAID has identified criteria for country selection based on a combination of factors that relate to country needs and country opportunities for water programming (see box). These criteria only relate to countries receiving USAID WASH support.

USAID will use these WASH criteria to achieve a greater concentration of resources in fewer countries and fewer program areas. This approach is in line with the goals of the PPD-6 towards greater selectivity and focus to facilitate greater country-level impact and results.

Country Categories

USAID has established three categories for WASH focus: transformative impact, leveraged impact, and strategic priority. These three categories will be used as a framework to analyze funding and program priorities for WASH funds as part of the CDCS process.

Transformative Impact countries are determined based on an analysis of need, opportunity, and a favorable programming environment to significantly leverage USAID's resources. These include countries where there is an opportunity to have trans-formative impact on national level policies and to leverage host country resources for the development and implementation of services. These countries would have the requisite infrastructure, governance, track record, and expertise necessary to bring WASH programs to scale.

- Ethiopia is an example of a country that could meet the criteria as a transformative impact country. It has the requisite infrastructure, governance, and institutional expertise for USAID water programs to have a transformative impact. USAID is currently working with the Government of Ethiopia to increase access to improve water supply in rural communities and to achieve universal access to sanitation.

Leveraged Impact countries are based on the strategic application of limited USAID resources, and generally are countries in which relatively small investment levels are likely to generate significant impact in at least one dimension of WASH – such as first time access to an improved drinking water source, first time access to an improved sanitation facility, ending open defecation, or new hand washing practices. In such countries USAID will concentrate relatively small investment levels to generate significant impact on at least one dimension of WASH.

- An example of a leveraged impact program is the Philippines Water Revolving Fund (PWRF). USAID established the PWRF, in partnership with the Japanese International Cooperation Agency, to enhance financing options and opportunities for small water and wastewater utilities. With an approximately $5 million investment in technical assistance to the sector, the program was able to mobilize $93 million worth of loans ($56 million of which came from private commercial banks), for water supply and sanitation projects reaching two million people with new or improved access to basic water and wastewater services.

Strategic Priority countries or locations are places in which USAID anticipates continued WASH programs due to a combination of strategic considerations and development needs. These countries or locations include but are not limited to programs located in Lebanon, Jordan, Yemen, and West Bank-Gaza, and often feature large-scale programs with a focus on infrastructure development.

- For example, the USAID/West Bank and Gaza Mission recently completed the Dura Cluster Water Storage and Distribution System Project, which has provided approximately 43,000 Palestinians with reliable access to potable water for the first time. The $17.5 million water project consists of more than 12 kilometers of transmission pipelines, 92 kilometers of distribution networks to rural communities, and the construction of three large, concrete water-storage tanks, which have the combined capacity to hold 4,000 cubic meters of water.

These three categories, transformative, leveraged, and strategic priority, will be used to review budget priorities and to determine expected impacts from WASH programs during the development of a CDCS. When using Development Assistance (DA) funds, strong consideration should be given to providing transformative impact countries with additional resources to take programs to scale, while leveraged impact countries will need to ensure that their relatively smaller investments create appreciable and significant impact. USAID expects a mixture of DA funds to go toward both transformative and leveraged impact countries and locations. Economic Support Funds (ESF) are expected to continue to support WASH programs in strategic priority countries and locations. Missions will need to work closely with their Regional Bureaus, during the CDCS process when possible, to clearly define programming rationale using these categories.

In addition, the Strategy anticipates that USAID will continue to commit significant WASH funds to countries outside of the three-tier system through Disaster Risk Reduction (DRR), International Disaster Assistance (IDA) and emergency and non-emergency Food for Peace Title II funding. Emergency situations will require flexibility for rapid response in crises.

Projections for Wash Programming

Based on previously requested funding levels, USAID projects providing a minimum of 10 million persons with sustainable access to improved water supply and 6 million persons with sustainable access to improved sanitation over the next five years. It is expected that a vast majority of the results from USAID water programming will represent longer-term development programs in transformative impact and leveraged impact countries.

The implementation of the Strategy will begin immediately, and starting in FY 2014, The USAID Global Water Coordinator will coordinate an agency-wide effort to recommend funding levels of WASH priority countries based on priorities identified through the CDCS process and on the criteria established in this Strategy.[20]

The projected number of beneficiaries to be reached includes two distinct categories of impact – "sustainable first time access to water supply and sanitation services for unserved populations" and "improved access to sustainable water supply and sanitation services." SO 1 of this Strategy aims to increase both sustainable first time access to water and sanitation and to improve the sustainable water supply and sanitation to acceptable levels of service.

This projection is not a definitive target. The projection is based on estimating methods using available data, and this Strategy sets no commitment to reach a fixed number of people for any specific region or field Mission.

Among the quantitative and qualitative information gaps that persist are insufficient data to separate out first time access to safe water from those gaining access to improved quality of water, as well as disaggregated unit cost data for water and sanitation programs across all regions.

Historical USAID Water Funding and Beneficiaries Reached (Funding in Millions of USD)

	2008			2009			2010		
		Beneficiaries			Beneficiaries			Beneficiaries	
	Funding	Water Supply	Sanitation	Funding	Water Supply	Sanitation	Funding	Water Supply	Sanitation
Sub-Saharan Africa	174	1,195,570	579,521	165	1,490,911	359,587	191	1,346,057	2,054,859
Asia Pacific	69	2,221,695	1,060,750	107	2,625,256	830,581	177	1,278,431	555,389
Central Programs	20	0	0	16	0	0	21	0	0
Europe and Eurasia	7	51,149	0	2	74,340	0	2	4,596	1,986
Latin America and Carribean	24	464,152	229,145	14	1,393,153	75,107	39	181,828	84,873
Middle East and North Africa	96	3,807,000	4,421,475	177	208,000	65,038	89	33,572	191,998
Other	0	0	0	1	0	0	2	0	0
TOTAL	**390**	**7,739,566**	**6,290,891**	**482**	**5,791,660**	**1,330,313**	**522**	**2,844,484**	**2,889,105**

	2011			Total 2008-2011			Average 2008-2011		
		Beneficiaries			Beneficiaries			Beneficiaries	
	Funding	Water Supply	Sanitation	Funding	Water Supply	Sanitation	Funding	Water Supply	Sanitation
Sub-Saharan Africa	186	1,176,021	548,745	715	5,208,559	3,542,712	179	1,302,140	885,678
Asia Pacific	52	1,516,705	604,881	406	7,642,087	3,051,601	101	1,910,522	762,900
Central Programs	15	60,514	40,465	72	60,514	40,465	18	15,129	10,116
Europe and Eurasia	2	50,193	0	13	180,278	1,986	3	45,070	497
Latin America and Carribean	18	140,086	116,586	94	2,179,219	505,711	24	544,805	126,428
Middle East and North Africa	86	880,982	614,589	449	4,929,554	5,293,100	112	1,232,389	1,323,275
Other	2	0	0	5	-	-	1	-	-
TOTAL	**360**	**3,824,501**	**1,925,266**	**1,754**	**20,200,211**	**12,435,575**	**438**	**5,050,053**	**3,108,894**

Therefore, an important learning objective of this Strategy is to further develop key data points and currently unknown factors affecting water programming. The Strategy intends for these projections to drive approaches leading to greater impact in WASH programming and to develop, by the end of the strategy period, a unit cost analysis across all regions.

The projection of providing 10 million persons with sustainable access to improved water supply and 6 million persons with sustainable access to improved sanitation only represents the direct impact from USAID-funded water programming. It does not include additional multiplier effects expected from partner organizations, other donors, foundations, non-profit organizations, and the private sector. When leveraging projected impacts from partner organizations and host country efforts to bring interventions to national scale, the expected total may be significantly higher.

To accomplish SO 1, three intermediate results have been identified that will guide the planning and implementation of this effort.

IR 1.1: Increase First Time Access to, and Improved Access to, Sustainable Water Supply Services

In the Strategy, increasing "sustainable first time access" refers to access to an improved water source that is gained by previously unserved populations. Improving the "quality of sustainable water supply services" refers to increasing existing access to and improving the quality of an already improved water supply.[21] For both first time access, and improved access, methods for ensuring the sustainability of water quality in the long term should be incorporated into program design. This may include developing monitoring systems to ensure that water quality, as well as water supply, is sustained at acceptable levels.

South-South Knowledge Sharing
Transforms Urban Water Systems

Sometimes the most effective drivers of change are local. This was the principle that drove USAID efforts to improve water and sanitation access in Asian cities through the regional network called Water Links. Of 1.7 billion people who live in Asian urban centers, 340 million lack access to safe water supply. Water Links built water sector capacity by facilitating partnerships between water operators in cities throughout Asia. Partnerships paired mentor operators with expertise in water access, water quality, and sanitation with recipient operators looking to improve and extend their service. Both mentor and recipient organizations benefited from targeted trainings and knowledge sharing. "There is nothing like learning from one's peers," said Arjun Thapan, Chairman of the Water Links Board of Trustees. Water Links has to date orchestrated 60 such south-south partnerships, which have brought reliable access to safe water to more than one million urban dwellers in Asia.[22]

The developing world's population with access to an improved source of drinking water increased by 16 percent from 1990 to 2010, and most regions around the world are on track to reach the goal for access to improved water by 2015. Sub-Saharan Africa, however, is not on track, even though nearly 273 million people gained access to an improved source of drinking water between 1990 and 2010. In sub-Saharan Africa, 39 percent of the population still did not have access to an improved source of drinking water in 2012. Globally, 81 percent (653 million) of people who are still without access to an improved drinking water source live in rural areas.[23] To meet this development need, the Strategy, in alignment with the Global Health Initiative, Feed the Future, and the Water for the Poor Act emphasizes increased programming and focus in sub-Saharan Africa.

Urban coverage is currently far ahead of that in rural areas, but challenges remain and Missions must consider the scale of required investment and potential for sustainability. As the world rapidly urbanizes, with urban populations expected to increase over 50 percent to 4.9 billion between 2008 and 2030, maintaining coverage levels in urban settings will create new development challenges.

To accelerate access to water, USAID should support, when adequate resources are available, decentralization of responsibilities. This requires capacity building of local governments to engage communities, mobilization of financing for both system expansion and operations and maintenance, and oversight of public and private sector service providers. Increased support to small-scale, private water suppliers will also increase service access, particularly in smaller towns, peri-urban neighborhoods, and rural communities relying on non-networked, decentralized systems. Support to water and sanitation entrepreneurs can increase coverage and generate income, while recognizing the need to balance between expanded coverage and the price differential between municipal and private water service, particularly for those that have no alternatives.

Capacity building to ensure the continued quality of water resources, environmental sustainability, and climate resilience of all investments will be supported, especially where competition for scarce water resources is increasing. By focusing on capacity building and leveraging local partners, programs can minimize overreliance on donors and bolster lasting sustainability.[24] Collaborative relationships with other multi-lateral and bi-lateral donors, with local and international banks, and with financing institutions to increase attention to small-and medium-scale financing should be strengthened.

IR 1.2: Increase First Time and Improved Sustainable Access to Improved Sanitation Services[25]

USAID will highlight scaling up and strengthening its sanitation programs as a special area of focus in the Strategy. This includes bringing a number of proven innovative and successful approaches to scale through Community Approaches to Total Sanitation (CATS). In addition, these efforts will focus on combining the promising approaches of Community-Led Total Sanitation (CLTS) and sanitation marketing to generate sanitation demand and strengthen the supply of sanitation products and services at scale, leading to improved health for people in rural and urban areas.[26]

USAID/Egypt: Water and Sanitation Services at Scale

Since the 1970s, USAID has supported the development of a sustainable water sector in Egypt. Beginning with emergent major infrastructure projects to address critical health issues in large urban centers and coastal cities, USAID sustained investment in the sector through smaller but critical infrastructure projects in provincial and secondary cities and small-scale simple projects in the poorest governorates in Upper Egypt. USAID-supported institutional strengthening activities and focused on helping local utilities develop their management capabilities and operations and maintenance capacity. USAID supported Egypt's water sector policy and regulatory reform, establishing the holding company for Water and Wastewater and establishing 24 regional utilities. As a result of this nearly $4 billion program, more than one quarter of Egypt's population has gained access to new and improved water supply or sanitation service and basic human health has improved dramatically.

A number of innovative and successful approaches have increased access to sanitation on a small scale. However, governments and the global community have not been able to scale-up and sustain these efforts. Moving forward, Missions must increase the proportion of water investments focused on sanitation services, as sanitation is critical to health and access lags far behind that of water. Sanitation is too often left out of programs, almost universally falling behind in terms of priorities in water programming. This Strategy encourages Missions to focus on sanitation services in both urban and rural settings. In countries that are off track to meet the MDG for sanitation, and where diarrheal disease and under-nutrition are prevalent, Missions must add sanitation as a key element of their water, health, and nutrition activities.

Initiatives should promote the three pillars of WASH programming: support for sanitation infrastructure, the promotion of behavior change, and creation of an enabling policy and institutional environment including robust financing and the development of local ordinances and regulations to stop open defecation and/or establish latrine building codes. These approaches generate sanitation demand and strengthen the supply of sanitation products and services. With limited sanitation funding, these approaches can be integrated to add value to nutrition, HIV/AIDS, governance, education, and maternal and child health programming.

Starting with a Sustainable WASH Formula

While constructing latrines and wells has been essential for improved public health in rural Afghanistan, it is not enough to create lasting impact. With 89 percent of rural Afghans defecating out in the open, changing long-held behaviors was the key to long-term health results. USAID's Sustainable Water Supply and Sanitation (SWSS) project mobilized communities to seek change and to build their own capacity to make it happen. Rather than using subsidies for latrine construction, SWSS trained women, religious and community leaders, and teachers in hygiene promotion. The program also trained select community members in CLTS to ensure ownership and sustain-ability at the community level. With 42,129 new or renovated latrines for more than 294,900 people, now communities not only have the hardware, they have the tools and the will to make real change for years to come.[27]

USAID Missions can promote community or public toilets in dense settlements, as well as sanitation facilities in institutional settings, such as schools, health facilities, markets, and transportation hubs. It is important that these efforts involve youth associations and women's groups, and encourages the construction of facilities that are accessible for people with disabilities and the elderly. To accelerate expansion of decentralized sanitation solutions in rural and peri-urban areas not connected to a sewerage network, Missions should focus on the entire sanitation hierarchy in developing sustainable sanitation services. Integrating supply-demand marketing approach can help address sustainability concerns of past sanitation efforts, especially those that focused on highly subsidized rural and peri-urban latrine construction. USAID sanitation programs will need to address how and where to dispose of concentrated sewage and sludge to not add to the environmental (including groundwater and downstream) impacts from poorly constructed or maintained septic tanks.

Opportunities for development and testing of improved, low-cost sanitation and waste management technologies should be pursued, as well as innovative management and financing approaches, to ensure sustainability and facilitate more rapid expansion of basic sanitation solutions.

Targeting priority countries in sub-Saharan Africa and South and South East Asia is key to the sanitation component of the Strategy.

IR 1.3: Increase Prevalence of Key Hygiene Behaviors

Building on its long-standing experience in promoting improved hygiene behaviors, the Strategy calls for USAID to promote three hygiene practices with the greatest demonstrated impact on health: (1) hand washing with soap at critical times, (2) safe disposal and management of excreta, and (3) improving household water storage, handling and treatment.[28] In addition, community-based approaches to behavior change, e.g., through health and agriculture extension services, will support communities transitioning from open defecation.

To accomplish sustained behavior change in hygiene, engagement with service providers will be required to ensure that a full range of products and services are available to respond to the consumer demands of both women and men. Promoting these behavior changes should draw on diverse approaches, including mass media communications, community mobilization, social marketing, promotion in schools and health facilities, and interpersonal contact and outreach. Integrating WASH into HIV care and support programs is essential, and a growing body of literature has demonstrated that WASH practices are particularly important in programs to reduce the impact of HIV and AIDS.[30]

Improving the capacity of local public and private sector actors to design, implement and evaluate locally appropriate and effective hygiene and sanitation promotion programs, and to create demand for products and services is crucial. This includes working with a broad range of providers of hygiene products and services, including local private sector actors, to provide household water treatment and storage technologies, soap, and other products to facilitate optimal hygiene behaviors. Supporting private sector networks that are already successfully promoting and selling products and services can help create demand for products and services.

Small Actions Leading to Large-Scale Change

When added together, small, doable actions can result in large-scale, lasting change. USAID's Hygiene Improvement Project (HIP) focuses hygiene behavior changes around hand washing with soap at critical times, safely disposing of feces, and household water treatment and safe storage. By integrating hygiene into multiple sectors, such as in schools and with people living with HIV/AIDS, the project delivered sustainable results to communities and households. Follow-on project WASHplus builds on HIP's successes by targeting schools and expanding to include nutrition and maternal and child health. Its Schools Promoting Learning Achievement through Sanitation and Health project in Zambia focuses on learning, and uses a framework for sustainability around cross-cutting issues, such as menstrual hygiene management and gender equity. [29]

Bringing Sanitation Access to Girls at Schools

In Somalia, cultural norms require absolute privacy for girls when using the latrines. Without proper access to sanitation facilities, the academic achievement of all children can suffer from the high truancy rates that result from illness. USAID's School Environment and Education Development for Somalia (SEEDS) project works to maintain a healthy environment within schools by improving WASH facilities, especially ensuring the separation of latrines for boys and girls. Rehabilitation and construction of latrines and hand washing stations under SEEDS has led to an overall increase of 32,337 new students, of whom 12,666 were girls, and access for 3,686 female students to improved WASH facilities and practices in FY 2011 alone.[31]

So 2 – Manage Water in Agriculture Sustainably and More Productively to Enhance Food Security

SO 2 Measures of Success 2013-2018

- Increase water use productivity and efficiency in rainfed areas
- Increase water use productivity and efficiency in irrigated agriculture systems

To ensure there are sufficient water resources available to simultaneously meet expanding food security needs while supporting an effective local water use balance, food producers must use water resources more effectively, economically, and in more environmentally sustainable ways. SO2 helps guide the Agency's Feed the Future investments in rainfed and irrigated agriculture with special attention given to rainfed agriculture.

Country selection and prioritization for water programming in agriculture will align with Feed the Future priority countries.

An integrated approach to enhance food security must consider the water productivity of the basin as a whole, and the water related tradeoffs and synergies between health needs and different types of productivity in both agricultural and natural food producing systems. This approach addresses the competing demands of multiple water users including households, agriculture, and industry. These principles are being incorporated in multi-year strategies developed by Missions with Feed the Future programs and are continually reinforced during implementation through technical support from USAID Washington.

Sufficient investments are critical for watershed management, as well as creative use of "natural" infrastructure to reduce erosion, mitigate floods, and ensure that ecosystem services are maintained and preserved for use by future generations. By stressing the notion that farming is an enterprise and effective water management is an investment input, USAID support can help producers take on the responsibility of covering recurring costs that can be accrued to water management. This includes labor costs associated with irrigating a crop, pumping costs, water user fees, and routine irrigation system maintenance costs. By placing a price on irrigation water, or other water distribution mechanisms, efficient allocation of water can be encouraged that reflects water's value and scarcity, and the need to build sustainable systems.

Water management affects both coastal and watershed regions, and there is an important role of aquaculture in food production and fish as a source of protein in diets. The Strategy, however, is focused on land-based food production as the area with the greatest potential for development impact given USAID's current resource investments. USAID, through Feed the Future and other activities, can still support fisheries and coastal areas outside the focus of this Strategy.

The private sector is a key and necessary ally for implementation and scale-up of the most promising farmer associations, agro-processing companies, local water associations, and buyers which increasingly understand water-related risks associated with their agricultural supply chains, though may be not yet aware of how climate change and other stresses can exacerbate these risks. Priority should also be given to non-point sources of pollution from agriculture and livestock activities to encourage investment in pollution control practices and technologies by farmers, fishers, investors, managers, and other stakeholders.

The Strategy supports USAID efforts to improve agricultural water management by promoting transparency and inclusiveness, paying particular attention to elevating the voice of women and their rights and address their specific and unique needs in this process. Opportunities for women to lead in water governance and to increase their technical and management capacities should be identified and expanded, where appropriate. In addition, decentralized water use governance structures, such as local water user associations, community-based organizations, producer groups, and other local organizations often provide a platform where women gain management and leadership experience.[32]

Finally, regardless of the amount of food produced and its availability, nutritional status can still be undermined by many factors, including high rates of diarrheal disease that reduce absorption of nutrients and cause under-nutrition and stunting in children. Close coordination of production activities with nutrition, domestic water supply, sanitation, and hygiene interventions, already described in SO1, is essential to ensure a healthy and properly nourished population.

IR 2.1. Improve the Efficiency and Sustainability of Food Production from Rainfed Agricultural Systems

The greatest and most cost-effective potential for crop yield increases are in rainfed areas. Half of the increase in the projected demand for water by 2025 could be met by increasing the efficiency of irrigation in these areas.[33] Almost 80 percent of agricultural land worldwide relies on rainfed production systems, and these systems provide 62 percent of the world's food. There is a wide regional variation in the percentage of rainfed cultivated land. For example, the percentage of rainfed cultivated land in sub-Saharan Africa is almost 95 percent, for Latin America it is 90 percent, for the Near East and North Africa it is less than 70 percent and for South Asia it is less than 60 percent. Variable rainfall, dry spells, and droughts make rain-fed farming a high-risk endeavor. [34] This is particularly true in Africa, where only about 14 percent of the area is relatively free of stress.[35] USAID water programming will address the specific conditions of each region.

Support for small-scale, often privately and farmer-owned, micro-irrigation as a point of investment has the potential to impact millions of farmers in sub-Saharan Africa and Asia and drive significant improvements in crop yields in smallholder agriculture. The majority of the Feed the Future investments to increase agricultural production are in areas where farmers rely on rainfed agriculture, with significant small-scale irrigation efforts ongoing in Tanzania, Ghana, Mali, Tajikistan, Haiti, and Ethiopia that include rehabilitation of defunct irrigation schemes.[37] In places like Cambodia, Ethiopia, and Kenya, where climate change is increasing the severity of dry conditions in the dry season, USAID is helping rainfed agricultural systems incorporate supplemental irrigation measures.

A Winning Watershed Formula Feeds Haiti

About 60 percent of Haiti's nine million people are farmers, but the country still imports more than 50 percent of its food. Farmers need help protecting their land against extreme weather, preventing soil erosion, and improving agricultural productivity. Through integrated Feed the Future activities across entire watersheds, USAID's Feed the Future West (formerly WINNER) program is training Haitians in modern farming techniques and providing them with access to innovative science and technological advances.

The result is that while the watersheds improve, production and incomes increase simultaneously. In 2012, increases in productivity yielded $9 million in gross margin for farmers."I have been a farmer for five years and I used to barely scrape by," said farmer Ronald Champagne."I am producing a lot more since I received assistance, and my products are of better quality." Average production of corn increased by 341 percent, beans by 100 percent, and rice by 58 percent.[36]

In many instances, the major obstacle to rainfed agriculture is not the absolute volume of rainfall, it is management of that rainfall. There is potential for increasing production in dry, sub-humid, and semi-arid rainfed regions around the world because approaches already exist which can double productivity in rain-fed agriculture.[39]

New scientific innovations and information technologies can also be effective. Cases in point include the use of drought or flood tolerant seed varieties such as "scuba" rice and hydro-meteorological and forecasting tools, e.g. RANET (Radio and Internet Technologies for the Communication of Weather and Climate Information for Rural Development).[40,41]

Low-Tech Solutions to Water Scarcity

With 80 percent of the labor force employed in agriculture and 51 percent of the population living on less than $1.25 a day, the importance of water to the food security and economic development of Mali cannot be overstated. The key to combating the challenges of climate change lies in understanding local conditions. Poor soil infiltration rates were plaguing Mali's rainfed farms, so USAID worked with farmers on ridge tillage to help increase the efficiency of rainwater capture for improved agricultural production. Ridge tillage has boosted cereal crop yields by 30 to 50 percent and reduced water runoff by 20 to 40 percent. This relatively simple intervention has had wide-ranging impacts on food security, ecosystem health, and livelihoods. Recharged local water tables allowed women to establish dry season gardens, and increased the productivity of high-value tree crops that provide nutritious food, oil, construction, and forage products.[38] On a regional scale, ridge tillage techniques have been replicated in Senegal and the Gambia.

IR 2.2: Improve the Efficiency and Sustainability of Food Production from Irrigated Agricultural Systems

USAID will focus on increasing irrigated agriculture in select countries, including expanding irrigation in a responsible, sustainable, and climate-resilient way. The most cost-effective investment for USAID in irrigated systems will be in improving the efficiency of existing irrigation systems and building the capacity to manage those systems like a business, whether through community based organizations or private investments. While irrigation can provide significant benefits, the investment costs are also large, and financing will be a challenge in many places. The environmental and social implications of expanding irrigation, including pumping aquifers dry and leaving drinking water saline and unusable, can be significant and must be managed appropriately.

USAID supports building the capacity of producer groups to manage irrigation systems as a business within viable value chains.[43] Absent such an approach, irrigation infrastructure has often fallen into disuse after the end of a project because benefits were too small and/or users were unable to cover recurring costs.

Weak management and poor farming practices can adversely affect water resources, and water-related problems can adversely affect food production, and the availability of safe water for drinking. Stripping land of its natural vegetation, for example, can result in increased flooding, erosion, hillside instability, and reduced water infiltration. Improved agronomic practices such as drought and flood-tolerant varieties, forecasting tools, conservation tillage, cover crops, integration of perennials, and other practices can help raise production, enrich soil organic matter, and ensure that more rainfall is absorbed and used for production.

Investments in Year-Round Farming Pay Off

In Nepal, 66 percent of households experience food shortages each year. Farmers, who make up more than two thirds of the workforce, struggle to maintain a subsistence income because they have no access to irrigation facilities and limited finances to invest in them. The Nepal Economic Agriculture and Trade (NEAT) program helps cut input costs and boost crop productivity by installing new and rehabilitating existing irrigation systems and training local technicians to maintain them, which enables year-round farming. Sita Rana Magar, farmer and matriarch of a family of five said her output had increased tenfold:"I have no debt to pay and no problem of living hand-to-mouth." NEAT's irrigation team works closely with farmers to understand local water resources, water demands, and topography to determine and install appropriate, sustainable systems for specific areas. To foster ownership, farmer group members pool funds for maintenance, and individual farmers pay for their own operating costs.[42]

The following governance and management approaches can increase the efficiency of an irrigation system:

- Canal rehabilitation and maintenance;
- More efficient on-farm water management techniques;
- Incorporation of water-efficient and drought-tolerant crops and varieties;
- Combined use of surface water and groundwater;

- Adoption of innovative irrigation management schemes by water users associations; and
- Expanded reuse of treated wastewater.

In addition to multi-stakeholder decision-making and planning to ensure sustainability, a key step in moving toward more rational water management is to support policy initiatives that use pricing mechanisms to create incentives for more efficient water use and ensure fair and equitable water distribution, including a measurement system that disaggregates the impacts by gender. USAID will support efforts to carry out economic valuation of water and watershed ecosystem services to better account for the financial value of natural systems through cost-benefit analysis and impact assessments for Feed the Future programs.

The Strategy calls for programs to reduce the vulnerability of farmers to rainfall variability by diversifying household economies and investing in systems that are less vulnerable. The Strategy also supports increased efforts to encourage national and local stakeholders to incorporate climate and ecological vulnerabilities into their planning and decision-making processes.

4. Operational Principles

The Strategy's operational principles provide the foundation of how water issues will be integrated into USAID programming. These principles, consistent with the USAID Policy Framework, are essential to improving health and increasing food security programming.

Operational Principles:

- Support host country ownership
- Build in sustainability from the start
- Apply integrated approaches to development
- Leverage "solution holders" and partner strategically
- Promote gender equality and female empowerment
- Leverage science and technology
- Measure and evaluate impact
- Achieve resilience

Support Host Country Ownership

USAID should be responsive to each country's development path, in accordance with the principle of country-led development. Programs should seek out and support country leaders and institutions that can help promote water policies and programs in those countries. USAID investments should build on existing efforts, assessments, and programs that partner countries and/or other donors may have already taken, such as those undertaken in partnership with the Consultative Group on International Agricultural Research (CGIAR) centers, or collaboratively with U.S. universities.

Many nations are devolving responsibility for natural resource management, land use planning, water and sanitation service delivery, and economic zoning to regional or municipal governments. Due to a lack of attention to informal settlements and peri-urban areas, these populations may often be deliberately ignored or excluded from service by municipal administrations. At the same time, water challenges are inherently global in nature and some issues may be best dealt with at the multinational or trans-boundary level. Good practices in democratic governance emphasize the importance of working at the level where relevant decisions are made, with attention to the lowest and most local levels of government. USAID will invest in social analysis, capacity building and developing the capacity of institutions at the appropriate governance level. USAID will ensure that its programming is inclusive and transparent, and that planning processes involve a range of stakeholders, including men and women, youth, indigenous peoples and other vulnerable and marginalized populations.

Build in Sustainability from the Start

Approaches that support sustainable health and food security are indispensable to meet long-term development objectives. Globally, the water community is moving beyond a simple focus on supporting first-time water supply access to a more thoughtful consideration of ensuring that such access is sustained over time. USAID has been on the cutting edge of sustainable WASH service delivery in urban and rural areas, and this Strategy emphasizes our continuing commitment to this critical dimension of the water challenge. Sustainability of agriculture systems is the key to long-term food security. The pillars of sustainability for water programs include: integrated water resource management, sound governance and citizen participation, inclusion of the unique roles, responsibilities, needs, and access concerns of women, appropriate environmental design, and long-term behavior change.

Comprehensive Commitments Yield Access to Water

In Indonesia, where more than 100 million people lack access to safe water, and 61 percent of the urban population are not served by piped water, a strong relationship between USAID and the Government of Indonesia has been the key to the sweeping improvements in service delivery at all levels under the Environmental Services Program (ESP) and the Indonesia Urban Water, Sanitation, and Hygiene (IUWASH) project. Working side-by-side to improve the capacity of service providers to expand coverage, the involvement of civil-society in decision-making, and the enabling governance and financial systems, IUWASH is targeting sustainable, improved access for two million people and improved sanitation for 200,000.[44,45]

To support sustainability, the Strategy seeks to strengthen governance and institutional capacity building in such areas as adapting to the water-related impacts of climate change, delivering potable water to families, ensuring adequate supplies of water are available to support economic growth, and mitigating potential conflict around water scarcity at the local, national, and – with the U.S. Department of State – regional levels. This Strategy supports building the capacity of women to lead in water governance, calls for greater support for professional advanced degrees for local professionals in water-related fields, and emphasizes the development of technical and leadership capacities among local stakeholders.

Additionally, through the implementation of the Water and Development strategy, USAID will seek to invest in longer-term monitoring and evaluation of its water activities in order to assess sustainability beyond the typical USAID Program Cycle and to enable reasonable support to issues that arise post implementation.

Apply Integrated Approaches to Development

People depend on freshwater and coastal, marine ecosystems from "ridge to reef" to protect human health and safety, reduce risks from floods and droughts, support recreational and aesthetic functions, maintain water quality, and assimilate waste. Many of these functions are water and watershed related. Whether those critical services succeed or fail depends in a significant way on how the ecosystem water resources are managed. Policy tools, technical approaches, and market mechanisms that keep ecosystems and watersheds healthy will be integrated across water activities where their application supports USAID's development objectives in health and food security.

Integrated water resource management can help to ensure that demands for water are equitably met without significant negative impact on users and the environment. A comprehensive approach must consider the water productivity of the basin as a whole, and the water related tradeoffs and synergies between health needs and productivity in both agricultural and food producing, natural systems. An integrated approach addresses the competing demands of multiple water users -including households, communities, agriculture, and industry.

Leverage "Solution Holders" and Partner Strategically

Partnerships are a critical component to achieve the objectives and desired results of this Strategy. These partnerships will help develop innovative approaches to financing, specifically focused on achieving the Strategy's two core objectives.

Private sector partnerships should focus on promoting market-based models of service delivery to mobilize local entrepreneurs to provide water-related products and services along the entire value chain. The Strategy encourages coordination with new and existing partners and key solution-holders, particularly nongovernmental organizations, civil society, the private sector, local and national governments, and international donors. The Strategy supports a more concerted effort to encourage strategic relationships with bilateral and multilateral donors. Achieving the Strategy's goals will require concerted effort to leverage support through multilateral development banks and credit authority mechanisms, such as USAID's Development Credit Authority (DCA), the Export-Import Bank (EXIM), and the Overseas Private Investment Corporation (OPIC).

Promote Gender Equality and Female Empowerment

The Strategy ensures that programming interventions meet women's needs. This includes safety and security issues, specific concerns related to disaster risk reduction for women and children, reducing unpaid labor, and leveraging women's leadership. In both the home and the community, the role women play in ensuring the success of WASH programming and access to safe water and sanitation is paramount.

Women, in much of the developing world, are responsible for finding and providing water for their families and gardens. This is in addition to their responsibilities related to crop production and livestock care.

Women and girls also bear the significant time burden and risk of collection from distant water sources. Particularly in poor rural areas, improving access to safe water, and thereby reducing the time women and girls spend on collection can lead to increased girls' attendance at school and women's involvement in income generating community and leisure activities. Adequate sanitation facilities in schools (e.g., private, clean, and gender-segregated) are especially crucial for ensuring that girls remain in school.

The Strategy seeks to ensure that women are empowered to effectively advocate for their perspectives and priorities, and that they are able to fully engage as managers, and partners and entrepreneurs in water-related activities and enterprises. Missions can draw upon guidance provided in the Agency's Gender Equality and Female Empowerment Policy for integrating gender equality objectives in project design and gender advisors for technical support when developing and implementing water programs.

Wash for Life

USAID, with co-funding from the Bill & Melinda Gates Foundation supports promising new approaches in the WASH sector, through the WASH for Life initiative. Beginning in 2011, and continuing over four years, this $17 million partnership will use USAID's Development Innovation Ventures (DIV) program to identify, test, and help transition to scale evidence-based approaches for cost-effective and sustained WASH services in developing countries. The DIV WASH for Life program will contribute to the growing evidence base for cost-effective approaches in this sector and aims to attract further investment to those solutions that prove successful.

One innovative idea supported by DIV is a 12-month pilot by Sanergy to build and franchise a dense network of 60 low-cost latrines to residents in Lunga Lunga, a slum of Nairobi, Kenya. The program collects waste daily and processes it as fertilizer and biogas. Designed by MIT engineers and architects, the low-cost, modular hygienic latrines can be assembled in one day. The sanitation centers are franchised to local entrepreneurs and local youth groups. Revenue from the organic fertilizer and biogas energy add to the model's profitability. Sanergy aims to expand to 3,390 centers reaching 600,000 slum dwellers – creating jobs and profit, while aiming to reduce the incidence of diarrhea by 40 percent.[46]

Engaging women and Girls in Improvements in Sanitation and Hygiene

The Mara region is one of the driest areas in Kenya. Women must walk up to six hours to get water under dangerous conditions. USAID is partnering with World Vision International to improve women's status, empowerment and safety. The project, which will ultimately benefit more than 150,000 people, includes activities such as building wells and latrines near the community, programs to improve hygiene and sanitation practices, and capacity building for health workers. There has also been support for setting up water user associations and village savings and loan groups, emphasizing the participation and empowerment of women to take leadership positions that support the long-term sustainability of the new organizations.[48]

Leverage Science and Technology

When designed appropriately for the local context, and with attention to inclusiveness and gender sensitivity, innovations in science and technology can dramatically improve WASH and the management of water resources at regional, national, and local scales. Investments in technologies such as hydro-meteorological, and groundwater monitoring stations; ICT and geographic information systems; Supervisory Control and Data Acquisition (SCADA) systems and remote sensing; as examples, can help host governments better manage water and sanitation facilities, extend or improve access, promote better hygiene, protect crucial ecosystem services, and operate more transparently. The data collected through these sources can be used in hydrologic, hydraulic, and water management modeling software to predict the possible outcomes of water use strategies, assisting in a range of policy issues including transnational water agreements, national water use strategies, and municipal water management systems. These and other innovations such as conservation tillage, crop rotation, drought resistant crops, and new irrigation methods should lead to improved food security by increasing agricultural output, reducing costs, and building the resilience of local communities.

Achieve Resilience

Through increased efforts to support integrated water and watershed management practices, the Strategy seeks to strengthen adaptation and resilience to climate change, in accordance with the USAID Climate Change and Development Strategy. USAID's policy and program guidance on Building Resilience to Recurrent Crisis will further guide Agency programming and development approach.

Integrating Resilience and Wash to Tackle Poverty

While moving herds in search of pasture and water for their livestock herds and for their communities, pastoralists have been hit hard by changing weather patterns and decreased rainfall in Ethiopia. When they can better forecast these changes, they not only more readily adapt to changing climatic conditions but they can protect their livelihoods. USAID supports the integration of these needs with access to clean and sustainable water sources under the Water, Sanitation, and Hygiene Transformations for Enhanced Resilience (WATER) Project. This integration has led to greater resiliency through early warning systems and improvement of rangeland and water management through community mobilization and support to customary institutions, including peace building through a "do no harm" approach. So far, 146,000 people have access to water year round as a result of USAID assistance.[49]

Measure and Evaluate Impact

The Water Office within the Economic Growth, Education and Environment (E3) Bureau will establish a research agenda that defines key water and development challenges to be addressed as part of the Strategy implementation process and determine how impact will be measured and evaluated. This learning agenda will identify and disseminate lessons learned and best practices, including developing quantitative indicators and models related to the two SO. This should include collaborative research on the most effective integrated projects that combine water programs with other Agency program areas, such as global health, food security, conflict, education, and climate change. Both U.S. and host country institutions of higher education can be key partners in applied research to support the integration and wise use of water resources. Water-related results and lessons learned from Agency programs will be identified and widely shared within USAID and the broader international water community.

5. Organizational Roles and Responsibilities

This section highlights organizational elements necessary to ensure effective adoption of the Strategy's principles and priorities. This Strategy will have both long-term and short-term implications beginning in FY 2013 and continuing for the duration of this Strategy. The USAID Global Water Coordinator, working together with the leaders of the Bureau for E3, Bureau for Food Security (BFS), Bureau for Global Health (GH), Bureau for Policy, Planning, and Learning (PPL), selected offices in the Bureau for Democracy, Conflict and Humanitarian Assistance (DCHA), regional bureaus, and individual Missions will coordinate the implementation of the Strategy.

To promote effective Agency-wide coordination, a cross-cutting Water Sector Council will be established to bridge offices and responsibilities. The Council will serve as the technical platform for discussion and enhanced coordination on water-related topics at USAID. The Council will be chaired by the Global Water Coordinator and have representation from all regional and technical Bureaus and offices engaged in water-related programming.

The USAID offices and personnel listed below have specific roles related to water programming:

USAID Global Water Coordinator

The Coordinator is the senior representative within USAID responsible for coordinating the response to key water policy initiatives, including USAID's

Strategy. The Coordinator will also serve as the primary spokesperson and liaison with public and private organizations, including interagency partners such as the U.S. Departments of Defense and State, foreign development agencies, and Congress, to coordinate efforts.

Office of Water, Bureau for Economic Growth, Education, and Environment (E3/W)

The Office of Water, in the E3 Bureau, provides USAID with technical leadership in areas related to water. The Water Office is the principal USAID technical office responsible for implementation of this Strategy including providing technical support, developing appropriate indicators, and reporting progress toward the global projections. The Water Office provides Missions with technical support and training from Washington and in the field. The E3 offices of Global Climate Change and Energy and Infrastructure will also provide technical support and training.

Bureau for Global Health (GH)

The Bureau for Global Health's Environmental Health Team provides technical leadership for sanitation and hygiene within USAID, and is responsible for implementing water, sanitation and hygiene programs. In collaboration with the E3 Water Office, GH has worked to increase the visibility and quality of WASH programs in the Agency and to build Mission capacity to manage them. GH also has lead responsibility for implementing Strategy recommendations as part of maternal and child health, HIV/AIDS and nutrition programs.

Bureau for Food Security (BFS)

In order to increase food production, and therefore food security, the BFS is making significant investments in water, including irrigation, water for agro-processing, rain harvesting, and conservation of water resources through agro-forestry, among others. BFS is responsible for analyzing supply and demand for water as related to food security, and for ensuring that its investments to increase food production are sustainable. BFS will support the development of relevant tools, toolkits, training materials and best practices, in coordination with the Water Office, in areas related to the intersection of water, food production, food security, and the integration of health objectives into Feed the Future programs.

Bureau for Policy, Planning, and Learning (PPL)

PPL will ensure that the Strategy incorporates relevant policies and directives and is properly aligned with the CDCS process as well as other elements of the program cycle.

Office of Budget and Resource Management (BRM)

The BRM is responsible for USAID's internal budget process and thus plays a key role in budget formulation and allocation. Funding is provided in such a manner as to maximize development and humanitarian impact and comply with legislative directives.

Office of Conflict Management and Mitigation (CMM)

CMM is organized within DCHA to lead USAID's efforts to identify and analyze sources of conflict. CMM will work with officers developing water programs to analyze the role water can play in both conflict mitigation and prevention. CMM may also conduct conflict assessments for water programs and support the development of tool kits for programs in water conflict areas.

Office of U.S. Foreign Disaster Assistance (OFDA)

OFDA is charged by the President of the United States with directing and coordinating international U.S. Government disaster assistance. OFDA continuously monitors global hazards, identifies potential areas of need, and stands ready to respond whenever disaster strikes. Disasters related to water include floods and storms, emergency WASH needs, and responses to slow-onset emergencies such as prolonged drought leading to food insecurity.

Regional Bureaus

Regional bureaus work in concert with the E3 USAID's Global Water Coordinator, the E3 Water Office, and BRM to identify during the budget formulation and allocation processes where financial resources should be allocated. The technical experts in the regional bureaus are responsible for program design and evaluation, reviewing Mission CDCSs, and for determining if Missions in their region have allocated adequate funds to address Agency and country water-related development priorities and objectives.

Missions

Missions are responsible for programming water investments, managing for results, and for evaluating water projects in order to meet their development goals as articulated in their CDCS. Missions are also responsible for continuing to provide data through existing reporting processes to enable USAID to present a consolidated report to external stakeholders on USAID's overall water investments, as well as inventory best practices.

In budget requests and CDCS submissions, Missions proposing food security and global health programs should consider, as relevant and appropriate, how water interventions help support their program objectives. Missions proposing WASH programs should consider linkages between WASH programs, and food security and global health efforts.

CONCLUSION

Achieving water security for regions, nations and individuals is a great development challenge confronting the world today.

At the individual and community levels, nothing is more fundamental to a safer and more secure future than ensuring the health of each and every child in a family and in their community. Nothing is more important than providing those same children with access to clean and safe water, the chance to live in a healthy environment, and adequate nutrition. At a global level, this Strategy addresses the many challenges of sustainable access to safe water and sanitation services for all, as well as to improve food security, through the

Presidential Initiatives of Global Health, Feed the Future, and Global Climate Change.

The Strategy raises the importance and visibility of water as a development priority, and highlights its importance to meeting the development imperatives of improved health and increased food security. Improving human health and welfare, and maintaining the sustainability of natural systems, requires a coordinated global response to the challenges of water access for present and future generations. This Strategy reflects the commitment of the U.S. Government to work in partnership with the global community to meet these challenges.

End Notes

[1] USAID. Climate Change and Development – Clean Resilient Growth, 2012 http://transition.usaid.gov/our_work/policy_planning_and_learning/documents/GCCS.pdf.

[2] Food and Agriculture Organization (FAO),"Water Scarcity" 2012, http://www.fao.org/nr/water/topics_scarcity.html.

[3] World Health Organization, "Millennium Development Goal Drinking Water Target Met" 2012, http://www.who.int/mediacentre/news/releases/2012/drinking_water_20120306/en/.

[4] UNICEF and World Health Organization, *Progress on Drinking Water and Sanitation, Update 2* 2012, http://www.unicef.org/media/files/JMPreport2012.pdf.

[5] World Health Organization, *Water, Sanitation, and Hygiene Links to Health,* 2004, http://www.who.int/water_sanitation_health/factsfigures2005.pdf.

[6] Choffnes, Eileen R. and Mack, Alison, Rapporteurs, *Global Issues in Water, Sanitation, and Health: Workshop Summary,* Forum on Microbial Threats, Board on Global Health, Institute of Medicine of the National Academies. Washington, DC: National Academies Press, 2009, http://www.ncbi.nlm.nih.gov/books/NBK28462/pdf/TOC.pdf.

[7] Babcock, Jr., Roger W. and Walton Raymond, eds. Reston, VA: American Society of Civil Engineers, 2, 2008 USDA-ARS, Conservation and Production Research Laboratory: Colaizzi, P. D., Bliesner, R. D. and Hardy, L. A.,"A Review of Evolving Critical Priorities for Irrigated Agriculture," *Proceedings of The World Environmental & Water Resources Congress, May 12-16, 2008, Honolulu, Hawaii,* http://naldc.nal.usda.gov/download/37841/PDF.

[8] WWAP (World Water Assessment Programme), *The United Nations World Water Development Report 4: Managing Water under Uncertainty and Risk*, vol. 1 (Paris: UNESCO, 2012), 25, http://unesdoc.unesco.org/images/0021/002156/215644e.pdf.

[9] World Water Assessment Programme (page 46-47): http://unesdoc.unesco.org/images/0021/002156/215644e.pdf.

[10] See: http://transition.usaid.gov/resilience/.

[11] Intelligence Community Assessment, Global Water Security, ICA 2012-08, 2 February 2012.

[12] See: http://www.whitehouse.gov/sites/default/files/email-files/US_National_Action_Plan_on_Women_ Peace_and_Security.pdf.

[13] Bernauer, Thomas, Böhmelt, Tobias, Buhaug, Halvard, Gleditsch, Nils Petter, Tribaldos, Theresa, Weibust, Eivind Berg, and Wischnath, Gerdis, "Water-Related Intrastate Conflict

and Cooperation (WARICC): A New Event Dataset," *International Interactions: Empirical and Theoretical Research in International Relations*, 38 no. 4, (2012): 529-545; USAID, *Measuring Fragility: Indicators and Methods for Rating State Performance* (2005), http://pdf.usaid.gov/pdf_docs/PNADD462.pdf. See Bernauer and Kalbhenn International Water Cooperation and Conflict: A New Event Dataset, 2010. Center for Comparative and International Studies (CIS) and Institute for Environmental Decisions (IED), http://www.ib.ethz.ch/docs/currentpapers/IRCC_Datapaper_Main_copy.pdf. USAID. Measuring Fragility: Indicators and Methods for Rating State, 2005 http://pdf.usaid.gov/pdf_docs/PNADD462.pdf.

[14] These documents are the core documents indicating how the US government has raised the importance of development as a pillar of foreign policy. These are the guiding documents for reforms at USAID, and outline the path forward. For the PPD-6 see http://www.whitehouse. gov/the-press-office/2010/09/22/fact-sheet-us-global-development-policy. For the QDDR see: http://www.state.gov/s/dmr/qddr/index.htm. For USAID Forward see http://forward.usaid.gov. For the Policy framework see: http://transition.usaid. gov/policy/USAID_PolicyFramework.PDF.

[15] U.S. Agency for International Development/U.S. Department of State: *Addressing Water Challenges in the Developing World – A Framework for Action*: http://transition.usaid.gov/ our_work/cross-cutting_programs/water/docs/Framework_for_ Action.pdf.

[16] See remarks by Hillary Rodham Clinton on World Water Day 2010: http://www.state.gov/ secretary/rm/2010/03/138737.htm.

[17] See the Feed the Future website at: www.feedthefuture.gov/.

[18] See the Program Cycle overview at: http://pdf.usaid.gov/pdf_docs/PDACS774.pdf.

[19] See Water and Sanitation Program: Scaling Up Rural Sanitation Learning by Doing: Working at Scale in Ethiopia 2012, as an example of a transformative impact. http://www.wsp.org/ wsp/sites/wsp.org/files/publications/WSP-Ethiopia-at-scale.pdf.

[20] A CDCS is a five-year strategy (although it may be shorter for countries in transition) that focuses on USAID-implemented assistance and related USG non-assistance tools. See http://pdf.usaid.gov/pdf_docs/PDACS300.pdf.

[21] The MDG Joint Monitoring Program (JMP) defines an "improved" source of drinking water as one which adequately protects the source from outside contamination including piped water into dwelling, yard or plot; public tap or standpipe; borehole; protected dug well; protected spring; and rainwater collection."Unimproved" sources are defined as: surface water (river, dam, lake, pond, canal, irrigation channel); unprotected dug wells, and springs. USAID will employ the same definitions. See the JMP website: http://www.wssinfo.org/.

[22] See the Water Links website for more information: http://www.waterlinks.org/.

[23] UNICEF and World Health Organization (WHO 2012), 4-12.

[24] Examples include: Indonesia (see USAID, *Micro-Credit Finance of Water Connections to New PDAM Customers; Assessment Study, Cooperation Between PDAM Tanah Datar and Bank Rakyat Indonesia,* 2006, http://pdf.usaid.gov/pdf_docs/PNADH332.pdf; and National Institute of Urban Affairs," An Indo-US Programme to Develop Viable Urban Infrastructure Finance System," http://www.niua.org/fireD.asphttp://www.niua.org/fireD.asp, and Cambodia (see USAID, "The Cambodia MSME Project Website," http://www.cambodiamsme.org/.

[25] For purposes of this strategy, USAID uses the MDG JMP definition of improved access to involve the use of any of the following facilities: flush or pour flush to a piped sewer system, septic tank, or pit latrine; ventilated pit latrine; pit latrine with slab; or composting

toilet. Unimproved facilities include open pit latrines without a slab; buckets; shared facilities; and no facilities (i.e. bush or field). See: http://www.wssinfo.org/.

[26] See the Community-Led Total Sanitation website, www.communityledtotalsanitation.org/.

[27] For more information on SWSS, see the June 2012 issue of USAID's *Global Waters* magazine, http://transition.usaid.gov/our_work/cross-cutting_programs/water/globalwaters/gw_ezine_june_2012.html.

[28] UNICEF WASH Section, *Evidence base: Water, Sanitation and Hygiene Interventions:, Literature Review: December 2009* ;WASH Section, New York: UNICEF 2009.

[29] See USAID's Hygiene Improvement Project website: http://www.hip.watsan.net/ and USAID's WASHplus website: http://www.washplus.org/.

[30] USAID and World Health Organization, WHO. *How to Integrate Water, Sanitation and Hygiene into HIV Programmes,* Geneva: World Health Organization, 2010: http://pdf.usaid.gov/pdf_docs/PNADT495.pdf.

[31] For more information, see USAID/Somalia's "Making an Old School New: A Rehabilitated School Helps Usher in a New School Term", August 26, 2009: http://eastafrica.usaid.gov/en/USAID/Article/1235/Making_An_Old_School_New.

[32] This aspect of the strategy supports USAID's *Gender Equality and Female Empowerment* Policy 2012, see http://transition.usaid.gov/our_work/policy_planning_and_learning/documents/GenderEqualityPolicy.pdf.

[33] Seckler, David, Amarasinghe, Upali, Molden, David, de Silva, Radhika and Barker Randolph: *World Water Demand and Supply, 1990-2025: Scenarios and Issues,* Research Report 19 ,Colombo, Sri Lanka: International Water Management Institute, 1998, http://www.iwmi.cgiar.org/Publications/IWMI_Research_Reports/PDF/PUB019/REPORT19.PDF.

[35] Bationo, Andre, Hartemink, Alfred, Lungu, Obed, Naimi, Mustapha, Okoth, Peter, Smaling, Eric and Thiombiano, Lamourdia A., et.al. African Soils: Their Productivity and Profitability of Fertilizer Use, 2006. Background paper presented for the African Fertilizer Summit 9- 13th June 2006, Abuja, Nigeria 2006, 13.

[36] A. Gambrill."A Winning Program in Haiti Protects Lives and Livelihoods." *Global Waters*, USAID. Quarter I, 2012: http://transition.usaid.gov/our_work/cross-cutting_programs/water/globalwaters/gw_ezine_q1_2012.html.

[37] http://www.feedthefuture.gov/progress.

[38] "Feed the Future Helps Scale-Up Ridge Tillage in Mali." Feed the Future. February 28, 2012: http://www.feedthefuture.gov/article/feed-future-helps-scale-ridge-tillage-mali.

[39] Wani, Suhas P., Rockström, Johan and Oweis, Thieb eds., Molden, D. Karlberg, L. Rockström, J., and Falkenmark, M. "Water Resource Implications of Upgrading Rainfed Agriculture—Focus on Green and Blue Water Tradeoffs," *Rainfed Agriculture—Unlocking the Potential,* Comprehensive Assessment of Water Management in Agriculture, Series 7, Oxfordshire, UK and Cambridge, MA: CAB International, 2009, 44-53.

[40] DFID "Sowing the Seeds of Scuba Rice: How Flood-Resistant Rice is Helping South Asian Farmers Feed Their Families" 2010, http://www.dfid.gov.uk/stories/case-studies/2010/sowing-the-seeds-of-scuba-rice/.

[41] USAID and the U.S. National Oceanic and Atmospheric Administration (NOAA) have partnered to support the development and implementation of the RANET program designed to provide weather and seasonal forecasts to surrounding communities. RANET has expanded to some 16 African countries and work is being done for pilot programs in Asia and the Pacific See http://www.usaid.gov/results-data/success-stories/technology-helps-farmers-read-clouds and http://www.oar.noaa.gov/spotlite/archive/spot_ranet.html.

[42] For more information, see the USAID/Nepal NEAT webpage: http://nepal.usaid.gov/our-work/program-area/economic-growtheducation-a-food-security/402-nepal-economic-agriculture-and-trade-neat-activity.html.

[43] Malian farmers, for example, have been supported with FTF funding to better manage the Alatona irrigation scheme producing rice: http://transition.usaid.gov/ml/en/AEG/IICEM.html. Water User Associations are being created with USAID support in Khatlon province of Tajikistan to improve both production and business management of produce marketing, input access, and harvest storage options. See http://www.wuasp-car.com/index.php?module=ContentExpress&file=index&func=display&ceid=32&meid=9.

[44] USAID/Environmental Services Program, Final Report, December 2004-March 2010: http://pdf.usaid.gov/pdf_docs/PDACQ869.pdf.

[45] USAID/Indonesia,"USAID Indonesia Urban Water, Sanitation, and Hygiene Project (USAID IUWASH)": http://indonesia.usaid.gov/en/USAID/Activity/269/USAID_Indonesia_Urban_Water_Sanitation_and_Hygiene_Project_USAID_IUWASH.

[46] See the project description at: http://idea.usaid.gov/newsroom/articles/div/11-12-02-250.

[47] Leite, Marianna, After the Summit: Women's Access to Water and Policy Making in Brazil, (2010) *Gender & Development*, 18 ,2010,: 1, 69-79.

[48] See the Transboundary Water for Biodiversity and Human Health in the Mara River Basin, Kenya & Tanzania (TWB-MRB) webpage on the Global Water for Sustainability (GLOWS) Program website for additional information: http://www.globalwaters.net/projects/currentprojects/transboundary-water-for-biodiversity-and-human-health-in-the-mara-river-basin-kenya-tanzania-twb-mrb/.

[49] USAID/Ethiopia, "Water, Sanitation, and Hygiene Transformation for Enhanced Resilience (WATER)": http://ethiopia.usaid.gov/programs/global-health-initiative/projects/water-sanitation-and-hygiene-transformation-enhanced-resi.

In: U.S. International Water ...
Editor: Constance Alvarado

ISBN: 978-1-63485-066-7

Chapter 2

WATER AND SANITATION ASSISTANCE: USAID HAS INCREASED STRATEGIC FOCUS BUT SHOULD IMPROVE MONITORING*

United States Government Accountability Office

WHY GAO DID THIS STUDY

Millions of people in developing countries lack access to safe water and improved sanitation. Congress passed the Senator Paul Simon Water for the Poor Act of 2005 to improve access to safe water and sanitation for developing countries. In 2013, USAID released its first *Water and Development Strategy*, which includes the objective of improving health through sustainable WASH.

GAO was asked to review USAID's WASH efforts. Focusing on WASH activities in 9 selected countries, this report (1) describes recent activities and funding, (2) assesses USAID missions' efforts to plan and implement activities, and (3) assesses USAID's monitoring of activities. GAO selected a nongeneralizable sample of 9 countries from USAID's list of 22 priority WASH countries. These 9 countries received about 53 percent of funding attributed to WASH for fiscal years 2012 and 2013. GAO also selected 16 activities for detailed review in the 9 countries, primarily on the basis of levels of funding. GAO analyzed USAID WASH funding data for fiscal years 2012

* This is an edited, reformatted and augmented version of the United States Government Accountability Office publication, GAO-16-81, dated October 2015.

through 2014 and reviewed agency documents, interviewed mission officials, and visited sites in 2 African countries.

What GAO Recommends

GAO recommends that USAID take steps to improve monitoring and reporting of WASH activities, by identifying and addressing reasons for missions' inconsistent adherence with agency guidance. USAID generally concurred with the recommendations and, in particular, outlined steps it is taking to address the report's second recommendation.

What GAO Found

U.S. Agency for International Development (USAID) missions in the 9 countries GAO selected for its review reported implementing a variety of water supply, sanitation, and hygiene (WASH) activities in fiscal years 2012 through 2014. WASH activities included capacity building, behavior-change communication, infrastructure construction, technical assistance, policy and governance, and financing. The missions' funding for WASH activities in these countries ranged from $4.4 million to $53.4 million.

USAID missions in these 9 countries are taking steps to develop and implement plans for WASH activities, with some missions making more progress than others. These missions are also generally taking steps to address long-term sustainability when planning WASH activities, as directed by USAID guidance, including the *Water and Development Strategy*. USAID is in the process of developing additional guidance to help all its missions address the sustainability of WASH activities.

The completeness and accuracy of USAID's monitoring of WASH activities varied in the 9 selected countries. GAO found that, inconsistent with agency guidance, these missions did not (1) consistently set annual targets for 6 of 16 WASH activities, (2) disaggregate beneficiaries by gender for 6 of 10 water supply and sanitation activities, (3) verify the accuracy of beneficiary data for 3 of 10 water supply activities, and (4) report accurate numbers of beneficiaries for 6 of 8 sanitation activities. Mission officials cited a variety of reasons for adhering inconsistently with agency guidance in some instances and in others the reasons for inconsistent adherence were not clear. These limitations in the completeness and accuracy of monitoring information for WASH activities may inhibit the effectiveness of USAID's oversight of such activities and affect its ability to accurately report on progress in increasing access to safe water and sanitation.

ABBREVIATIONS

ADS	Automated Directives System
APHIA	AIDS, Population and Health Integrated Assistance
BCC	behavior-change communication
CB	capacity building
CLTS	community-led total sanitation
DRC	Democratic Republic of the Congo
F	financing
Field Guide	*USAID Water and Development Strategy Implementation Field Guide*
IC	infrastructure construction
PAD	project appraisal document
PG	policy and governance
SDSH	Santé pour le Développement et la Stabilité d'Haiti
2005 Act	Senator Paul Simon Water for the Poor Act of 2005
2014 Act	Senator Paul Simon Water for the World Act of 2014
State	Department of State
SUWASA	Sustainable Water and Sanitation in Africa
TA	technical assistance
UN United	Nations
UNICEF	United Nations Children's Fund
USAID	U.S. Agency for International Development
WASH	water supply, sanitation, and hygiene
WATER	Water, Sanitation, and Hygiene Transformation for Enhanced Resilience
Water Strategy	*USAID Water and Development Strategy, 2013-2018*
WHO	World Health Organization

October 6, 2015

The Honorable Karen Bass
Ranking Member
Subcommittee on Africa, Global Health,
Global Human Rights, and International Organizations
Committee on Foreign Affairs
House of Representatives

The Honorable Richard J. Durbin
United States Senate

The Honorable Earl Blumenauer
House of Representatives

The Honorable Ted Poe
House of Representatives

In many developing countries, a lack of access to safe water and basic sanitation severely limits economic growth and development and leads to suffering and death for millions each year.[1] In 2005, Congress passed the Senator Paul Simon Water for the Poor Act of 2005 (2005 Act), making access to safe water and basic sanitation for developing countries a U.S. foreign policy objective.[2] The 2005 Act required, among other things, that the Department of State (State), in consultation with the U.S. Agency for International Development (USAID), develop and implement a U.S. water and sanitation strategy that included designation of high-priority countries where need is greatest. In 2010, we reported on State's implementation of U.S. water and sanitation assistance.[3] In 2014, Congress passed the Senator Paul Simon Water for the World Act of 2014 (2014 Act) to strengthen implementation of the 2005 Act by improving the capacity of the U.S. government to implement, monitor, leverage, and evaluate programs to provide first-time or improved access to safe drinking water, sanitation, and hygiene to the world's poorest on an equitable and sustainable basis.[4] The United States has primarily provided water and sanitation assistance through USAID. In fiscal years 2005 through 2014, USAID reported allocating about $3.5 billion for water supply, sanitation, and hygiene (WASH).

You asked us to review the U.S. government's international water and sanitation assistance. This is the second of two reports responding to your request. In July 2014, we provided information about U.S. agencies' funding for international water-related assistance and the extent to which the agencies complied with congressional spending requirements.[5] In this report, we (1) describe the types of WASH activities USAID has implemented in selected countries and funding it has provided for these activities, (2) assess the extent to which USAID guidance has informed the agency's efforts to plan and implement WASH activities in these countries, and (3) assess USAID's monitoring and evaluation of selected WASH activities.

We focused our review on USAID's WASH activities in 9 selected countries: the Democratic Republic of the Congo (DRC), Ethiopia, Haiti, Indonesia, Kenya, Tanzania, Senegal, Uganda, and Zambia. We selected these countries from 22 that USAID designated as priorities for WASH assistance in fiscal year 2014.[6] The 9 selected countries accounted for $155 million—about 53 percent—of funding attributed to congressional spending requirements for WASH by all tier 1 and tier 2 priority countries in fiscal years 2012 and 2013.[7] For all three reporting objectives, we interviewed USAID and State officials and conducted semi-structured telephone interviews with officials at the USAID missions in the 9 selected countries. Because we judgmentally selected these 9 countries, our findings from these interviews cannot be generalized to all USAID missions. We also conducted fieldwork in Tanzania and Ethiopia to meet with U.S. and host country officials, implementing partners, and beneficiaries and to conduct site visits for USAID water and sanitation activities. To describe the types of WASH activities USAID has implemented and funding it has provided for these activities, we obtained funding data and descriptive information from each mission regarding its ongoing or planned WASH activities in fiscal years 2012 through 2014. To assess the reliability of the data and information we obtained, we reviewed documentation and interviewed agency officials to identify and correct any missing data and any errors. We determined that the data and information were sufficiently reliable to provide general information on funding and types of WASH activities.

To assess the extent to which USAID guidance informed planning and implementation of WASH activities in the 9 selected countries, we reviewed USAID guidance, mission-level WASH plans, and other documents.[8] We assessed the status of the USAID missions' efforts to develop WASH plans, steps they have taken to implement these plans and to incorporate USAID guidance into their recent or planned WASH activities, and steps they have taken to address sustainability. We conducted a detailed review of documents and information for 16 activities[9] that we selected on the basis of factors such as levels of funding and types of activities implemented in the 9 countries in fiscal years 2012 through 2014.[10]

To assess monitoring of selected WASH activities in the 9 countries, we reviewed award agreements and modifications, performance management plans, monitoring and evaluation plans, and monitoring reports for the 16 activities that we selected for detailed review. In addition, to assess USAID evaluations for activities in the 9 countries, we assessed the quality of 14 evaluations for WASH activities completed in fiscal years 2012 through 2014 against established principles of evaluation, which correspond with USAID's requirements for evaluations. We did not assess the selected USAID missions' compliance with the requirement to conduct evaluations. Appendix I provides additional information about our objectives, scope, and methodology.

We conducted this performance audit from July 2014 to October 2015 in accordance with generally accepted government auditing standards. Those standards require that we plan and perform the audit to obtain sufficient, appropriate evidence to provide a reasonable basis for our findings and conclusions based on our audit objectives. We believe that the evidence obtained provides a reasonable basis for our findings and conclusions based on our audit objectives.

BACKGROUND

Legislation and Key Agency Roles Related to WASH Assistance

The 2005 Act made access to safe water and sanitation a U.S. foreign policy objective. Specifically, the act aimed to promote the provision of access to safe water and sanitation to countries, locales, and people with greatest need, including the very poor, women, and other vulnerable populations.

Congress passed the 2014 Act to strengthen the 2005 Act by improving the capacity of the U.S. government to implement, monitor, and evaluate programs that increase access to safe water, sanitation, and hygiene. The 2014 Act requires, among other things, that USAID ensure that WASH projects are designed to promote maximum impact and long-term sustainability. The 2014

Act also calls for rigorous monitoring and evaluation to assess improvements in WASH.

State's Office of Conservation and Water within the Bureau of Oceans and International Environmental and Scientific Affairs is responsible for the development and implementation of U.S. foreign policy on international water and sanitation assistance. USAID's Office of Water within the Bureau for Economic Growth, Education and Environment is responsible for coordinating, managing, and overseeing USAID's response to water policy initiatives, including the 2014 Act.

Annual Spending Requirement for WASH Assistance

Annual appropriations acts funding foreign operations have established a spending requirement for WASH. These spending requirements have ranged from $315 million to $365 million since 2012.[11] USAID and State have provided annual guidance to USAID missions regarding activities for which they may attribute funding to the annual spending requirement. The guidance notes that for attribution, proposed activities must be able to demonstrate an impact on water supply, sanitation, or hygiene through objectively verifiable indicators to measure progress.[12] The 2005 Act required State to submit an annual report to Congress detailing the status of WASH efforts; these reports included information from USAID about funds allocated to WASH activities to meet the spending requirements for WASH. This reporting requirement was repealed by the 2014 Act.

USAID Water and Development Strategy

Improving health outcomes through the provision of sustainable WASH is the first strategic objective of *USAID's Water and Development Strategy, 2013-2018 (Water Strategy).*[13] To achieve this objective, the Water Strategy includes three key WASH-related goals:[14]

- **Increase first-time and improved access to sustainable water supply.** First-time access refers to access to an improved water source that is gained by previously unserved populations.[15] Improved access refers to enhancing existing access to, and the quality of, an already improved water supply.

- **Increase first-time and improved access to sustainable sanitation.** First-time access to improved sanitation generally refers to access to a pit latrine with a slab, septic system, or similar type of improved sanitary facility.[16] Improved access to sanitation generally refers to improvement of an existing sanitation facility.[17]
- **Increase adoption of key hygiene behaviors.** The Water Strategy recommends the promotion of three hygiene practices with the greatest demonstrated impact on health: (1) hand washing with soap at critical times; (2) safe disposal and management of excreta; and (3) improving household water storage, handling, and treatment.

Table 1. Standard and Custom Indicators That USAID Recommends to Track Its Water Supply, Sanitation, and Hygiene (WASH) Goals

WASH goal	USAID-recommended standard and custom indicators[a]
Increase first-time and improved access to sustainable water supply	• ***Number of people gaining access to an improved drinking water source*** • **Number of people receiving improved service quality from existing improved drinking water sources** • Percentage of households using an improved drinking water source • Number of policies, laws, agreements, regulations, or investment agreements (public or private) that promote access to improved water supply and sanitation • Public sector expenditures on drinking water and sanitation as a percentage of national budget • Percentage of a drinking water utility's supply that is non-revenue • Percentage of population using an improved drinking water source • Percentage of children under five who had diarrhea in the past 2 weeks
Increase first-time and improved access to sustainable sanitation	• ***Number of people gaining access to an improved sanitation facility*** • **Percentage of households using an improved sanitation facility** • *Number of improved toilets in institutional settings* • Number of policies, laws, agreements, regulations, or investment agreements (public or private) that promote access to improved water • Public sector expenditures on drinking water and sanitation as a percentage of national budget • Percentage of population using an improved sanitation facility • Percentage of children under five who had diarrhea in the past 2 weeks • Percentage of population in target areas practicing open defecation • *Number of communities certified as "open defecation free"* • Number of individuals trained to implement improved sanitation methods[b]
Increase adoption of key hygiene behaviors	• *Number of liters of drinking water disinfected with point-of-use treatment products as a result of U.S. government assistance* • Percentage of households in target areas practicing correct use of recommended household water treatment technologies • Number of households with soap and water at a hand-washing station commonly used by family members in U.S. government assistance programsb

Legend:

Boldface type marks indicators that the U.S. Agency for International Development (USAID) intends to use to track its progress in meeting its Water and Development Strategy's first strategic objective.

Italic type marks indicators that we used to assess USAID's monitoring of WASH activities.

Source: GAO presentation of USAID indicators in the Water and Development Strategy Implementation Field Guide (March 2014). | GAO-16-81

Notes: The Department of State (State) and USAID jointly developed standard indicators to measure the results of U.S. foreign assistance. While these indicators do not reflect all indicators tracked by individual bureaus, offices, and missions across State and USAID on an ongoing basis, this standard set of indicators allows for the consolidation of certain key results to provide a picture of what is being achieved with foreign assistance resources to Congress and the public. Performance target and result data are to be collected against these indicators on an annual basis and reported to State's Office of U.S. Foreign Assistance Resources.

[a] Indicators listed are standard indicators unless otherwise noted For this review, if a selected activity's indicators did not include any of the WASH indicators listed in USAID's *Water and Development Strategy Implementation Field Guide*, we selected at least one of the activity's indicators related to drinking water, sanitation, and hygiene efforts, as applicable.

[b] USAID's *Water and Development Strategy Implementation Field Guide* recommended the use of this custom indicator.

Unimproved drinking water sources	Improved drinking water sources
Surface sources: River, dam, lake, pond, stream, canal, irrigation channels **Other sources:** Unprotected dug well, unprotected spring, cart with small tank/drum, bottled water	**Piped water on premises:** Piped household water connection in user's dwelling, plot, or yard **Other sources:** Public taps or standpipes, tube wells or boreholes, protected dug wells, protected springs, rainwater collection

Unimproved sanitation facilities	Improved sanitation facilities
Open defecation: Disposal of human feces in fields, forests, bushes, open bodies of water, beaches, or other open spaces **Unimproved facilities:** Pit latrines without slab or platform, hanging latrines, and bucket latrines with no mechanism to ensure hygienic separation of human excreta from human contact **Shared facilities:** Sanitation facilities of an otherwise acceptable type that are shared between two or more households	**Improved facilities:** Facilities that are likely to ensure hygienic separation of human excreta from human contact and are not shared or public. For example: • Flush/pour flush to - piped sewer system - septic tank - pit latrine • Ventilated improved pit latrine • Pit latrine with slab

Source: GAO presentation of World Health Organization and United Nations Children's Fund information. | GAO-16-81

Note: USAID adheres to the World Health Organization's and the United Nations Children's Fund's definitions for access to drinking water and sanitation.

Figure 1. USAID Definitions Related to Access to Drinking Water and Sanitation.

Figure 1 shows USAID's definitions for improved and unimproved drinking water and sanitation.

In March 2014, USAID issued the *Water Strategy Implementation Field Guide* (Field Guide) as a tool to help missions understand and implement the Water Strategy. The Field Guide requires that USAID missions track progress toward the three WASH-related goals using standard indicators and some custom indicators, as shown in table 1.

State's annual reports to Congress detailing the status of WASH projects have included country-level results for two of the standard WASH indicators: (1) number of people gaining access to an improved drinking water source and (2) number of people gaining access to an improved sanitation facility.[18] This information has also been reported in USAID's annual report on international water-related assistance, *Safeguarding the World's Water*.

The Water Strategy projected that during the next 5 years, at least 10 million persons would receive sustainable access to improved water supply and 6 million persons would receive sustainable access to improved sanitation.[19] These projections included persons receiving first- time and improved access to water supply and sanitation.

According to the Water Strategy and the Field Guide, to achieve the greatest impact, WASH projects should include the following elements:

- expanded access to "hardware" (e.g., water and sanitation infrastructure and hygiene commodities);
- required "software" activities to promote behavior changes for sustained improvements in water and sanitation access/service and hygiene practices; and
- an improved enabling policy and institutional environment, including strengthened financial frameworks and public-private partnerships for WASH.

The Field Guide states that the level of effort in each area may vary depending on local context and other factors. In addition, the Water Strategy and the Field Guide highlight gender issues in the water sector as a key focus. They note that the burden of inadequate access to water and sanitation often falls heavily on women and girls and that WASH activities should promote gender equality and female empowerment to address the needs and opportunities of both men and women.[20]

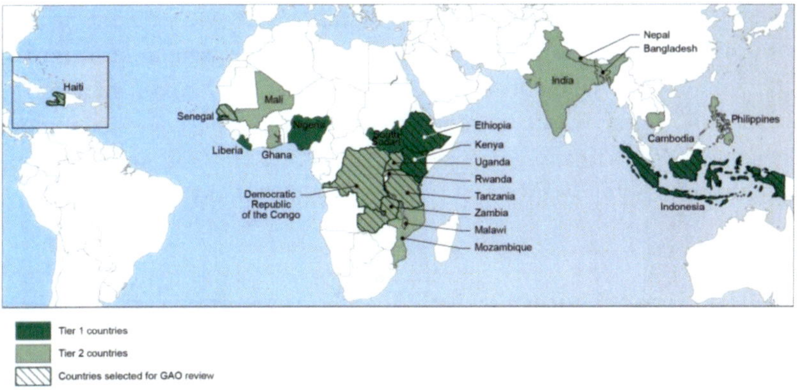

Source: GAO analysis of U.S. Agency for International Development (USAID) data; Map Resources (map). | GAO-16-91

Note: According to USAID, Tier 1 countries are those where the agency found an opportunity to have a transformative impact on national-level policies and to leverage host country resources for the development and implementation of assistance. Tier 2 countries are those where USAID determined that relatively small investment levels were likely to generate a significant impact in at least one dimension of WASH. USAID includes the Sahel region—consisting of parts of Burkina Faso, Cameroon, Chad, the Gambia, Mali, Mauritania, Niger, Nigeria, and Senegal—in its list of tier 2 priority countries, in addition to the countries shown above.

Figure 2. USAID-Designated Tier 1 and Tier 2 Priority Countries for Water Supply, Sanitation, and Hygiene (WASH) Assistance, Nine of Which GAO Selected for Review.

WASH Sustainability

The Water Strategy emphasizes the importance of sustainable WASH services. USAID policy defines sustainable services as public services in which host country partners and beneficiaries take ownership of development processes and maintain project results beyond the life of a USAID project.[21] The strategy notes that the pillars of sustainability for water projects include integrated water resource management, sound governance, and appropriate environmental design, among other factors. In particular, regarding projects to provide first-time and improved access to quality water supply services, the strategy states that methods for ensuring the sustainability of water quality in the long term should be incorporated into project design and that this may include developing monitoring systems to ensure that water quality and supply are sustained at acceptable levels. Further, regarding sanitation services, the strategy notes that it supports the development and testing of improved, low-cost sanitation and waste management technologies, as well as innovative management and financing approaches, to ensure sustainability and to facilitate more rapid expansion of basic sanitation solutions.

USAID Priority Countries for WASH Assistance

In fiscal year 2014, USAID designated 22 countries as tier 1 or tier 2 priorities for WASH assistance on the basis of their need and the opportunity to achieve significant impact.[22] The 6 tier 1 countries are those where USAID found an opportunity to have a transformative impact on national-level policies and to leverage host country resources for the development and implementation of assistance. The 16 tier 2 countries are those where USAID determined that relatively small investment levels were likely to generate a significant impact in at least one dimension of WASH. USAID reported attributing about $463.1 million to the annual spending requirements for WASH in fiscal years 2012 through 2014 for these priority countries. Figure 2 shows the locations of the tier 1 and 2 priority countries.

USAID Missions' WASH Activities and Funding Varied

In the nine countries we selected for our review, USAID missions reported $214.5 million in WASH funding for 74 activities, with 6 key focus areas.[23] Each mission's total funding for WASH activities in these nine countries ranged from about $53.4 million, in Indonesia, to about $4.4 million, in Haiti.[24] Missions reported implementing from as many as 19 WASH activities, in Kenya, to as few as 3 activities, in both Tanzania and Zambia.[25] The most frequent key focus areas for these activities were capacity building and behavior-change communication, followed by infrastructure construction, technical assistance, policy and governance, and financing.[26] Figure 3 provides additional detail about the activities implemented by each selected mission in fiscal years 2012 through 2014.

Table 2 describes the 74 activities' key focus areas and provides examples of the types of WASH activities in each focus area for the nine selected countries.

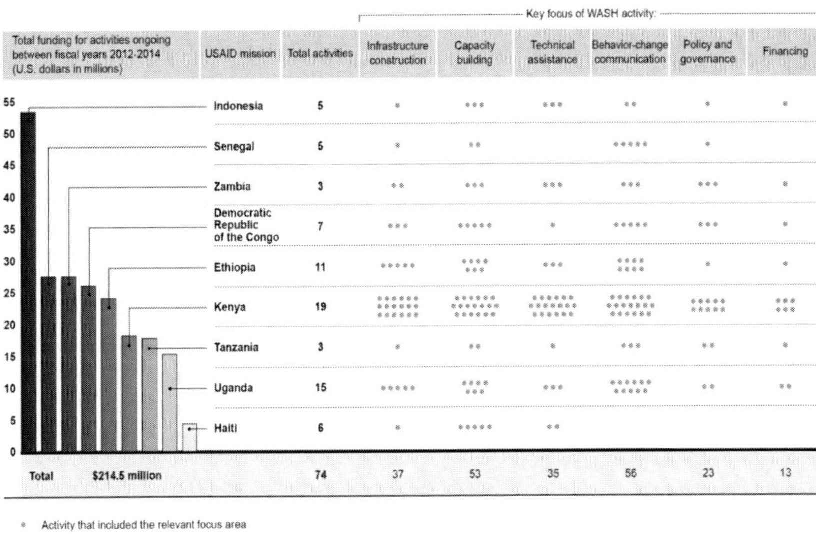

Source: GAO analysis of U.S. Agency for International Development (USAID) data. | GAO-16-81

Notes: Each mission's total funding generally represents allocations for WASH components of activities that were ongoing at the mission at any time in fiscal years 2012 through 2014, including, when applicable, WASH funding for years before fiscal year 2012 and through fiscal year 2014. We included obligations in cases where USAID was not able to provide allocations for WASH activities, such as when a USAID mission obligated previously unplanned funding to an activity for WASH. Some activities started before fiscal year 2012, and some were ongoing at the end of fiscal year 2014.

Missions reported multiple focus areas for most activities.

Figure 3. Funding and Focus Areas of Water Supply, Sanitation, and Hygiene (WASH) Activities Implemented by Nine USAID Missions in Fiscal Years 2012-2014.

Table 2. Key Focus Areas and Examples of Selected Water Supply, Sanitation, and Hygiene (WASH) Activities Implemented in Nine Selected Countries in Fiscal Years 2012-2014

Key focus area	Description	Example
Behavior-change communication	Communication strategies to promote positive WASH behaviors	Promoting healthy WASH behaviors, including hand washing and water purification, through social marketing campaigns
Capacity building	Development of partner organizations' or governments' capacity related to WASH	Training local governments and community organizations in managing, operating, and maintaining water systems
Financing	Increasing the availability and accessibility of financing for WASH services	Promoting partnerships between water service providers and microfinance banks to increase access to water and sanitation services
Infrastructure construction	Construction of new systems or rehabilitation of existing systems for drinking water and sanitation	Constructing community water points and latrines in households and schools
Policy and governance	Support of an enabling environment to facilitate WASH delivery	Supporting host governments in developing pro- WASH policy, legal, and regulatory frameworks, instruments, and institutions
Technical assistance	Provision of expertise for achieving WASH program objectives	Assisting local governments in developing plans and budgets for WASH activities

Source: GAO analysis of U.S. Agency for International Development (USAID) information. | GAO-16-81

Note: We identified key focus areas for 74 water supply, sanitation, and hygiene activities implemented by USAID missions in, respectively, the Democratic Republic of the Congo, Ethiopia, Haiti, Indonesia, Kenya, Tanzania, Senegal, Uganda, and Zambia. Missions reported multiple focus areas for most activities.

Missions noted that some of the 74 reported activities included WASH as one of several other components, such as maternal-child health, nutrition, or natural resources management. For example, the Senegal mission's Yaajeende Agriculture activity is an agricultural activity that aims to increase nutritional status by diversifying foods produced and eaten. The activity includes a WASH component to expand access to clean drinking water and improved sanitation and also includes water resources management efforts to promote effective small irrigation technologies. In another example, the DRC mission's Integrated Health Project aims to improve the enabling environment and increase the availability of services, products, and practices for family planning; maternal, newborn, and child health; nutrition, malaria, and tuberculosis; and WASH in targeted health zones.

Table 3. Geographic Focus of Selected USAID Water Supply, Sanitation, and Hygiene Activities Implemented in Nine Selected Countries in Fiscal Years 2012-2014

Geographic focus area	Number of activities	Total funding (U.S. dollars in millions)
Rural, peri-urban, urban	38[a]	$62.4
Rural and peri-urban	12	$69.5
Rural	10	$22.4
Urban and peri-urban	7	$6.3
Urban	7	$54

Source: GAO analysis of U.S. Agency for International Development (USAID) information. | GAO-16-81

Notes: Funding shown generally represents allocations for water supply, sanitation, and hygiene (WASH) components of 74 activities implemented by USAID missions in, respectively, the Democratic Republic of the Congo, Ethiopia, Haiti, Indonesia, Kenya, Tanzania, Senegal, Uganda, and Zambia and includes activity funding through fiscal year 2014. We included obligations in cases for which USAID was not able to provide allocations for WASH activities, such as when a USAID mission had obligated previously unplanned funding to an activity for WASH. According to a USAID official, USAID missions do not employ a universal definition for rural, peri-urban, and urban, and missions are encouraged to defer to national census authorities' definitions of these terms.

[a] For one of the 38 activities, USAID reported targeting rural and urban areas but not peri-urban areas (e.g., small towns).

Additionally, the nine missions reported implementing WASH activities targeting three types of geographic areas: rural, peri-urban (e.g., small towns), and urban.[27] For example, the Tanzania mission's Integrated Water, Sanitation, and Hygiene activity targeted rural and peri-urban areas (e.g., small towns) to provide support including community piped water schemes, rehabilitated wells, training for community groups to operate and maintain water systems, and school latrines, among other efforts. In contrast, the Indonesia mission's Indonesia Urban Water, Sanitation, and Hygiene activity aimed to increase access to clean water and improve sanitation facilities for people in urban settings, where about half of Indonesia's population lives. According to the Indonesia mission, the activity focused on fostering demand for WASH, building capacity, and strengthening the policy and financing environment for WASH through its work with central and local governments, including entities responsible for delivering water and sanitation services.

Table 3 includes information about the geographic focus of the 74 activities in the nine selected countries.

Appendix II provides additional information about WASH activities implemented by USAID missions in each of the nine countries.

USAID Is Taking Steps to Enhance Strategic WASH Approach and Address WASH Sustainability

In the nine countries we selected for our review, we found that USAID missions are taking steps to develop and implement WASH plans and are incorporating the Water Strategy's principles and approach into recent and planned WASH activities. The missions are considering sustainability as part of WASH project planning, and USAID's Office of Water is developing guidance for missions specific to WASH sustainability.

Missions in Selected Countries have Taken Steps to Develop and Implement WASH Plans and Enhance Strategic WASH Approach

We found that the missions in the nine selected countries have made varying degrees of progress in developing WASH plans. In addition, we found that these missions have generally taken steps to implement WASH plans and enhance the strategic approach in recent WASH projects.

Our review of documents describing these missions' strategic approach to WASH, as well as interviews with officials at the nine missions, found that five of the missions had completed WASH plans. Missions' WASH plans may consist of one or more project appraisal documents (PAD), which document a project's design and expected results.[28] The remaining four missions were in the process of developing or finalizing such plans.

- Five of the nine missions—in Ethiopia, Indonesia, Kenya, Tanzania, and Zambia, respectively—had completed WASH plans and taken steps to implement these plans as of July 2015. The Indonesia and Tanzania missions initially developed WASH plans in fiscal year 2009.29 The Zambia, Kenya, and Ethiopia missions completed WASH plans more recently, in the period from fiscal year 2013 to fiscal year 2015.
- Three of the nine missions—in the DRC, Senegal, and Uganda, respectively—were in the process of finalizing WASH plans as of July 2015.
- One mission, in Haiti, had begun developing a plan for a more strategic approach to its WASH activities as of February 2015. According to mission officials, the mission had focused its recent water-related activities on agriculture, water resources management, and water productivity. In response to the Water Strategy, the mission has begun to develop a plan for future WASH activities that will consider findings from a fiscal year 2015 mission WASH sector assessment.

Our review of missions' WASH project documents, as well as semi-structured interviews with officials at the nine missions, also found that these missions have generally taken steps to implement WASH plans and incorporate the Water Strategy's strategic principles and approach in recent WASH projects, such as by implementing activities to support expanded access to water and sanitation infrastructure along with activities to support capacity building and behavior change. Although four of the nine missions acknowledged that their prior WASH projects lacked strategic focus, all nine missions cited ongoing efforts to develop more focused, strategic, and impactful WASH projects in accordance with the Water Strategy.

Figure 4 summarizes our findings regarding the status of the nine selected missions' WASH plans and their recent and planned WASH activities as of July 2015. Appendix II provides additional information about the selected missions' WASH activities.

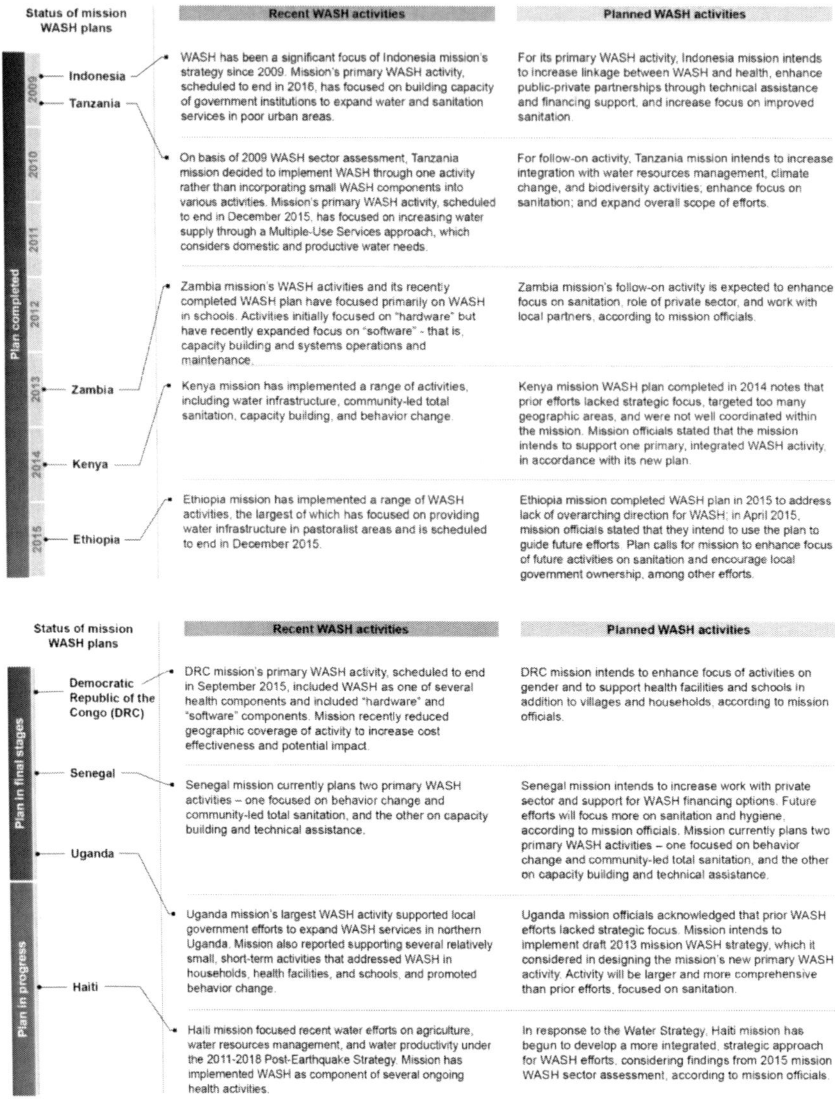

Source: GAO analysis of U.S. Agency for International Development (USAID) data and information. | GAO-16-81

Note: Years shown are fiscal unless otherwise noted. We reviewed WASH activities that the missions implemented in fiscal years 2012 through 2014 as well as activities that the missions were planning during that period.

Figure 4. Status of Nine USAID Missions' Water Supply, Sanitation, and Hygiene (WASH) Plans and Recent and Planned Activities as of July 2015.

Missions Are Considering Sustainability as Part of WASH Planning

Our review of mission documents and information showed that the nine selected missions are generally taking steps to address sustainability as part of their WASH project planning, as described below:[30]

- WASH PADs for all five of the missions that had completed these documents—in Indonesia, Kenya, Tanzania, Uganda, and Zambia—included sustainability analyses.[31] For example, the Kenya mission's WASH PAD included an annex with a two-page sustainability analysis that described the mission's planned steps to address challenges to sustainable WASH, including challenges related to choice of technologies and technical approaches, weaknesses in governance, long-term financing needs, and the need for sustained behavior changes. The Uganda and Zambia missions' PADs sustainability analyses did not address WASH specifically but described sustainability considerations in the context of broader project efforts (e.g., health or education).
- Missions' award documents for the 16 activities we reviewed in detail generally described sustainability considerations. For example, the award agreement for a WASH activity in Senegal noted that decades of development experience have shown that merely building water and sanitation infrastructure is not sufficient to deliver adequate service or to ensure that such services are sustainable. The agreement added that WASH efforts would incorporate principles such as optimal balance between hardware and software activities, local ownership and decentralized management of WASH infrastructure and service delivery, and use of appropriate and affordable technologies.[32]

- Five of the nine missions conducted more in-depth WASH sustainability assessments to inform planning and design of WASH activities. The Ethiopia mission completed two studies between April 2012 and February 2013 to assess existing water supply schemes and potential strategies to improve sustainability for the mission's primary ongoing WASH activity, which was implemented in several regions.[33] In addition, in 2013 and 2014, USAID piloted a tool that it developed with Rotary International to provide an in-depth assessment of the sustainability of several WASH activities implemented by missions in three of the nine countries we selected for our review—Indonesia, Kenya, and Tanzania.[34] The assessments included an overview of the WASH sector in each country; a description of steps taken to apply the sustainability tool; results of using the tool, including scores for institutional, management, financial, technical, and environmental sustainability; key findings; and priority areas for action and recommendations by intervention category (i.e., water supply, sanitation, or hygiene).[35] A USAID official noted that for this type of tool to be effective, USAID would need to use it systematically and incorporate a feedback loop to ensure that results are incorporated into future activities. The Indonesia mission incorporated the results of using the tool into its latest WASH plan, while the Kenya and Tanzania missions completed their latest WASH plans before the results of the tool were available. However, Kenya and Tanzania mission officials stated that they are considering results of the assessments in planning their upcoming activities.

Mission officials in the selected countries generally told us that they plan to enhance the focus of future activities on sustainability. For example, officials at the Senegal mission noted that they conducted a sustainability analysis in the process of developing the mission's forthcoming PAD related to WASH. To enhance its focus on sustainability, the mission is planning, among other efforts, to implement private sector reform activities and more directly engage with the government of Senegal to better ensure ownership of water and sanitation projects. Mission officials in the DRC said that they are in the process of determining the cost of monitoring all of their WASH activities after implementation is completed to better assess sustainability. The officials noted that UNICEF, which is implementing one of the mission's WASH activities, is using its own resources to monitor the activity to assess sustainability 1 year after completion and that the mission would like to standardize the process for other activities.

USAID's Office of Water Is Developing Guidance Specific to WASH Sustainability

USAID's Office of Water has begun to develop guidance to address WASH sustainability. The March 2014 Field Guide states that USAID's Office of Water will develop guidance for missions to program effectively for sustainable WASH services, including sustainability indicators, monitoring options, and tools for assessing sustainability.[36] To inform future WASH sustainability guidance, in December 2014, USAID's Office of Water developed a draft technical paper that included recommendations for designing projects to achieve sustainable WASH service and for confirming sustained results through longer-term monitoring.[37] The draft paper described, among other things, the types of indicators that could be used to assess sustainability, including the functionality and reliability of WASH services. Additionally, the draft paper recommended that USAID's Water Office should

- supplement its sustainability guidance with questions and resources for missions to consider in WASH project design,
- adopt sustainability indicators to assess the functionality and reliability of WASH services,
- develop an approach to post-project monitoring, and
- conduct an evaluation on the sustainability of WASH services in a sampling of countries.

According to USAID officials, USAID's Office of Water intends to finalize its guidance for addressing the sustainability of WASH activities in 2015. The officials stated that the guidance will likely be broad and include a menu of options, tools, and resources that can be tailored to the context in which each mission implements WASH activities.

USAID mission	Activity	Total WASH funding[a]	Annual targets for key WASH indicators identified	Annual performance data for key WASH indicators reported	Performance data on water and sanitation were disaggregated by gender
Democratic Republic of the Congo	Integrated Health Project	$13,377,369	Yes	Yes	No
	Sustainable WASH Interventions: Healthy Villages Program in Two Health Zones, Provinces of Katanga and Kasai Oriental	$9,141,101	Yes	Yes	Yes
Ethiopia	Water, Sanitation, and Hygiene Transformation for Enhanced Resilience	$10,984,723	No[b]	Partial[c]	Yes
	Your Health is in Your Hands Water, Sanitation and Hygiene program	$4,300,000	Yes	Yes	No
Haiti	Services de Santé de Qualité pour Haïti	$1,324,487	Yes	Yes	N/A
Indonesia	Indonesia Urban Water, Sanitation, and Hygiene	$38,696,403	Yes	Yes	No[d]
	Urban Household Water Supply in Indonesia (Water Hibah)	$10,000,000	Yes	Yes	N/A
Kenya	AIDS, Population and Health Integrated Assistance Program (APHIA) Plus Northern Arid Lands Service Delivery	$3,024,887	No[e]	Yes	No
	Sustainable Water and Sanitation in Africa Project to Support Urban WASH in Kenya (Kenya 2)	$500,000	Yes	Yes	No
Senegal	USAID/Senegal Millennium Water and Sanitation Program (PEPAM)	$20,866,000	Partial[f]	Yes	No
	Health Communication and Promotion Program Component	$700,000	Yes	Yes	N/A
Tanzania	Tanzania Integrated Water, Sanitation and Hygiene Program	$17,443,586	Yes	Yes	Yes
Uganda	WASHPlus: Supportive Environments for Healthy Communities	$500,000	No	No	N/A
	Health Marketing Initiative (AFFORD II)	$450,000	No	No[g]	N/A
Zambia	Schools Promoting Learning Achievement through Sanitation and Hygiene	$15,200,000	Partial[h]	Yes	Yes
	Partnership for Integrated Social Marketing	$6,900,000	Yes[i]	Yes	N/A
Total		$153,408,556	Yes = 10 No = 4 Partial = 2	Yes = 13 No = 2 Partial = 1	Yes = 4 No = 6 N/A = 6

Legend: Yes; No; Partial; N/A Not applicable

Source: GAO analysis of U.S. Agency for International Development (USAID) data. | GAO-16-81

Notes: USAID ADS ch. 203 (revised in November 2012) requires that at the start of each activity, the implementer establish a monitoring plan with indicators and associated targets to assess progress on the activity. This policy also requires that all people-level indicators at mission strategy, project, or activity level be sex-disaggregated to help USAID assess contributions to gender equality and female empowerment.

In this report, "key WASH indicators" refers to WASH indicators whose use is recommended in USAID's *Water Strategy Implementation Field Guide* (Field Guide) (March 2014). These indicators include number of people gaining access to an improved drinking water source, number of people gaining access to an improved sanitation facility, and number of liters of drinking water disinfected with point-of-use treatment products. If an activity did not use WASH indicators recommended in the Field Guide, we selected at least one indicator related to water supply, sanitation, and hygiene efforts, as applicable.

[a] Total funding generally represents allocations for WASH components of activities through fiscal year 2014. We included obligations in cases for which USAID was not able to provide allocations for WASH activities such as when a USAID mission obligated previously unplanned funding to an activity for WASH.

[b] This activity had final targets—for example, "about 203,000 to 213,000 beneficiaries gaining access to an improved drinking water source." However, the activity lacked annual targets, which are necessary to monitor progress on an ongoing basis.

[c] Annual performance data were reported for the activity's water indicator but not for the sanitation and hygiene indicators.

[d] According to officials at the Indonesia mission, the unit of measurement for this activity was households, which cannot be disaggregated by gender. The implementer estimated that the number of beneficiaries on the basis that each household had an average of 5 people.

[e] This activity lacked annual targets for key WASH indicators for the remaining 2 years of its 5-year duration.

[f] Targets were included for the activity's water and sanitation indicators but not for hygiene indicators for liters of water disinfected and villages attaining "open defecation free" status.

[g] The monitoring plan for this activity did not identify any WASH-specific indicators.

[h] Although the monitoring plan for this activity did not identify annual targets for key WASH indicators, the mission-wide annual reports included targets for two school WASH activities until fiscal year 2013 and targets specific to Schools Promoting Learning Achievement through Sanitation and Hygiene for fiscal years 2014 and 2015.

[I] Although the monitoring plan for this activity did not identify annual targets for key WASH indicators, the mission-wide annual reports included targets related to this activity.

Figure 5. Nine Missions' Compliance with USAID Requirements in Documenting Annual Targets and Reporting Performance Data for Selected Water Supply, Sanitation, and Hygiene (WASH) Activities Implemented in Fiscal Years 2012-2014.

MONITORING AND REPORTING FOR USAID'S WASH ACTIVITIES IN SELECTED COUNTRIES HAD LIMITATIONS, ALTHOUGH EVALUATIONS WERE GENERALLY SOUND

Our detailed review of 16 WASH activities in the nine selected countries found limitations in the monitoring and reporting of some activities' performance, although most evaluations we reviewed were sufficiently reliable and methodologically sound for their intended purposes.[38] Monitoring plans for 6 activities did not consistently include annual targets for key WASH indicators in accordance with USAID requirements, limiting the missions' ability to measure progress toward WASH goals.[39] Also, while the annual reports submitted by the 16 activities' implementers generally included performance data for the key WASH indicators, the reports for 6 activities did not present data disaggregated by gender as required by USAID policy. Moreover, contrary to agency guidance, missions did not verify beneficiaries for at least 3 activities aimed at increasing access to improved drinking water sources and overstated beneficiaries for 6 activities aimed at increasing access to improved sanitation facilities, calling into the question the accuracy of USAID's annual reporting about progress toward these WASH goals. The reasons for the missions' inconsistent adherence to agency guidance regarding annual targets, gender-disaggregated data, and verification of beneficiaries were generally unclear, while USAID officials provided differing reasons for the inaccurate reporting for sanitation activities. In contrast, 12 of the 14 performance evaluations that we reviewed were sufficiently reliable and methodologically sound for their intended purposes.

Monitoring Plans and Reports for Several Activities Lacked Annual Targets, and Reports did not Consistently Disaggregate Beneficiaries by Gender

For more than one-third of the 16 activities we reviewed, the monitoring plans did not consistently include annual targets for key WASH indicators, such as the number of people gaining access to an improved drinking water source. USAID requires that at the start of each activity, the implementer establish a monitoring plan with indicators and associated targets to assess progress on an activity.[40] Additionally, State guidance to USAID missions requires that proposed activities demonstrate impact through objectively verifiable indicators to measure progress toward WASH goals, if funds are attributed to the annual congressional spending requirement for international water and sanitation assistance. We found that, of the 16 activities, 3 lacked annual targets for key WASH indicators for the entire duration of the activity and 1 lacked annual targets for key WASH indicators for the remaining 2 years of the activity's 5-year duration. Additionally, 1 activity included annual targets for the drinking water and sanitation component but not for the hygiene component; for another activity, annual targets were clearly identified for 2 years of its 5- year duration.

While annual reports for the 16 activities we reviewed generally included performance data for key WASH indicators,[41] the reports for more than half of activities with beneficiary indicators did not disaggregate data for these indicators by gender as USAID policy requires.[42] Ten activities had indicators to measure numbers of beneficiaries gaining increased access to an improved water source or improved sanitation facility. However, for 6 of these activities, the missions did not disaggregate performance data by beneficiaries' gender, making it difficult for USAID to assess the activity's contributions to gender equality and female empowerment.

Reasons for the inconsistency that we observed in the selected missions' adherence to agency guidance regarding annual targets and gender-disaggregated data were unclear. USAID officials in Washington, D.C., noted the absence of consistent reasons among the nine missions for a lack of regular reporting on annual targets and of gender-disaggregated data. Officials at the DRC mission noted that staffing constraints had generally limited their ability to monitor their WASH activities and that they planned to hire additional staff to improve monitoring.

Figure 5 summarizes our findings regarding the nine selected missions' compliance with USAID requirements in documenting targets and reporting performance data for the 16 activities that we reviewed.

Missions Did Not Verify Results for Some Water Activities and Overstated Results for Most Sanitation Activities

For some of the 10 activities with a water component that we reviewed, the missions did not verify the numbers of beneficiaries as required, and for most of the 8 activities with a sanitation component, the missions overstated the numbers of beneficiaries.[43] As a result, because activity performance data contribute to USAID's annual public reporting of WASH results, the data that USAID uses to report the numbers of people gaining access to an improved drinking water source and to an improved sanitation facility annually may not be accurate. The reasons for lack of verification for water beneficiaries were generally unclear, and mission officials, as discussed later, generally provided varying reasons for the inaccurate reporting for sanitation activities. Figure 6 summarizes our findings regarding the nine selected USAID missions' compliance with State and USAID requirements for verifying beneficiaries of water activities and reporting beneficiaries of sanitation facilities.

USAID mission	Activity	Total WASH funding[a]	USAID mission verified beneficiaries gaining access to improved drinking water sources	USAID mission's reporting of beneficiaries gaining access to improved sanitation facilities was consistent with State and USAID guidance
Democratic Republic of the Congo	Integrated Health Project	$13,377,369	No	No
	Sustainable WASH Interventions: Healthy Villages Program in Two Health Zones, Provinces of Katanga and Kasai Oriental	$9,141,101	No	Yes[b]
Ethiopia	Water, Sanitation, and Hygiene Transformation for Enhanced Resilience	$10,984,723	Yes	N/A[c]
	Your Health is in Your Hands Water, Sanitation and Hygiene program	$4,300,000	Not clear	No
Haiti	Services de Santé de Qualité pour Haïti	$1,324,487	N/A	N/A
Indonesia	Indonesia Urban Water, Sanitation, and Hygiene	$38,696,403	Yes	No
	Urban Household Water Supply in Indonesia (Water Hibah)	$10,000,000	N/A[d]	N/A
Kenya	AIDS, Population and Health Integrated Assistance Program (APHIA) Plus Northern Arid Lands Service Delivery	$3,024,887	Not clear[e]	No
	Sustainable Water and Sanitation in Africa Project to Support Urban WASH in Kenya (Kenya 2)	$500,000	Not clear	N/A
Senegal	USAID/Senegal Millennium Water and Sanitation Program (PEPAM)	$20,866,000	Not clear[f]	Yes
	Health Communication and Promotion Program Component	$700,000	N/A	N/A
Tanzania	Tanzania Integrated Water, Sanitation and Hygiene Program	$17,443,586	Yes	No
Uganda	WASHPlus: Supportive Environments for Healthy Communities	$500,000	N/A	N/A
	Health Marketing Initiative (AFFORD II)	$450,000	N/A	N/A
Zambia	Schools Promoting Learning Achievement through Sanitation and Hygiene	$15,200,000	No[g]	No
	Partnership for Integrated Social Marketing	$6,900,000	N/A	N/A
Total		$153,408,556	Yes = 3 No = 3 Not clear = 4 N/A = 6	Yes = 2 No = 6 N/A = 8

Legend: Yes; No; N/A Not applicable; State Department of State; USAID U.S. Agency for International Development; WASH Water Supply, Sanitation, and Hygiene

Source: GAO analysis of USAID data. | GAO-16-81

Notes: State and USAID guidance on standard indicators, including indicator definitions, require that on completion of construction or rehabilitation of an improved water source, the implementer or evaluator estimate beneficiaries by verifying usage through observations and interviews with initial users and by taking into account the time to collect water and the quantity of water produced by the water source. USAID follows World Health Organization and United Nations Children's Fund definitions of improved water source and improved sanitation facility, which state that a pit latrine without a slab or platform is an "unimproved" sanitation facility. Additionally, only facilities that are not shared or are not public are considered improved.

[a] Total funding generally represents allocations for WASH components of activities through fiscal year 2014. We included obligations in cases for which USAID was not able to provide allocations for WASH activities, such as when a USAID mission obligated unplanned funding to an activity for WASH.

[b] Although the DRC mission's Sustainable WASH Interventions activity adhered to the definition of improved sanitation facilities, USAID mission officials could not verify the accuracy of the reported data.

[c] The Ethiopia mission's Water, Sanitation, and Hygiene Transformation for Enhanced Resilience activity included a sanitation component but used a custom indicator,

rather than the standard indicator for number of people gaining access to improved sanitation facilities, to assess progress.

[d] The Indonesia mission's Water Hibah activity included verification of households that obtained a new water connection from the local utility. Because the activity used a custom indicator (number of new household water connections) rather than a standard indicator, we designated it as not applicable to State's and USAID's requirement for verification of the standard indicator of people gaining access to an improved drinking water source. USAID guidance gives missions flexibility to report using custom indicators relevant to their activities.

[e] Officials at the Kenya mission stated that they had conducted a data quality assessment in November 2013. However, we found an inconsistency in beneficiary data in the monitoring report for fiscal year 2014. One section of the report stated that 116,779 beneficiaries had gained access to an improved drinking water source, while another section of the report stated that 55,200 beneficiaries had gained access to the improved source. The mission officials acknowledged this error and stated that the lower estimate was more accurate.

[f] Although officials at the Senegal mission stated that they undertook regular site visits and data quality assessments to verify estimated beneficiaries, they did not provide documentation for such verification.

[g] According to officials at the Zambia mission, while the beneficiaries gaining access to improved drinking water in schools could be verified, the beneficiaries from communities surrounding the school, who also have access to water points constructed or rehabilitated under this activity and are included in reported results, could not be verified.

Figure 6. Nine USAID Missions' Compliance with State and USAID Requirements for Verifying Beneficiaries of Water Activities and Reporting Beneficiaries of Sanitation Activities.

Lack of Verification of Beneficiaries Gaining Access to Improved Water Sources

For 7 of the 10 activities that we reviewed, the missions and implementers did not undertake efforts to verify reported beneficiaries or did not document such efforts, calling into question the accuracy of the reported results. State and USAID guidance for the standard indicator of access to an improved drinking water source requires that the implementer or evaluator verify these estimates by assessing factors such as the amount of time the user spent in collecting water and the quantity of water produced by the new or rehabilitated water source.[44] The reasons for the lack of verification, or documentation of verification, of beneficiaries of the 7 activities were generally unclear.

Documents for only 3 of the 10 activities—in Tanzania, Ethiopia, and Indonesia, respectively—evidenced verification efforts and, in one case, corrective actions related to verifying results of activities to provide access to improved drinking water.[45] For 4 other activities, it was unclear, on the basis of documents that the missions provided and mission officials' responses to our queries, whether the reported results were verified. For example, although officials at the Senegal mission stated that they undertook regular site visits and data quality assessments to verify estimated beneficiaries, they did not provide documentation of such verification. Moreover, the midterm evaluation for this activity noted that estimating the number of beneficiaries on the basis of the Senegalese government's assumptions was feasible at the planning stage but not during implementation, when the number of beneficiaries could be verified. The evaluation also noted that, absent precise numbers of beneficiaries, it was difficult to determine the cost-effectiveness of the investment.[46]

For the remaining 3 activities—2 in the DRC and 1 in Zambia—mission officials informed us that they had not verified the reported results and cited varying reasons.

- In the DRC, annual reports for 2 of the WASH activities we reviewed indicated that more than 900,000 people gained access to an improved drinking water source between fiscal years 2012 and 2014.[47] In USAID's annual report for fiscal year 2013, the number of beneficiaries reported as gaining access to an improved drinking water source in the DRC constituted more than 13 percent of worldwide beneficiaries (446,989 out of 3,509,090). Officials at the DRC mission informed us that they had been unable to conduct a data quality assessment planned for October 2014 because of challenges that included security concerns and difficulties of traveling in a country with limited roads. Mission officials also noted that estimating numbers of people who gained access to an improved drinking water source was complicated by the displacement of population in the areas where USAID was implementing its WASH activities.
- In Zambia, the mission reported in fiscal years 2012 and 2013 that 82,606 and 62,098 people, respectively, gained access to an improved drinking water source as a result of the mission's two school WASH activities. Officials at the Zambia mission informed us that although they undertook verification of data for the ongoing school WASH activity's sanitation component, they did not verify all reported beneficiaries of the water component, which included the school population as well as people from the surrounding communities. The

officials noted that while beneficiaries gaining access to improved drinking water in schools could be verified, it was not possible to verify beneficiaries from communities surrounding the school, who also have access to water points constructed or rehabilitated under this activity.

Inaccurate Reporting of Beneficiaries Gaining Access to Improved Sanitation Facilities

For six of the eight activities aimed at increasing access to improved sanitation facilities that we reviewed, the activities' implementers reported beneficiaries for facilities that did not meet USAID's definition of an improved facility.[48] As a result, the data reported to track progress toward the goal of increasing access to improved sanitation are likely overstated. USAID uses WHO and UNICEF definitions of an improved sanitation facility, which state that a pit latrine without a slab or platform is an unimproved sanitation facility and that only facilities that are not shared or are not public are considered improved.[49]

For six of the eight sanitation activities that we reviewed, implementers tracked and reported numbers of people gaining access to sanitation facilities that included unimproved latrines and shared facilities.[50] Mission officials generally provided differing reasons for the inaccurate reporting for these activities, such as perceived agency emphasis on reporting beneficiaries and adherence to host-government policy or practice. USAID officials in Washington, D.C., noted that, to some extent, the missions' inconsistency in accurately reporting on the sanitation indicator resulted from inadequate understanding of USAID's definition of improved sanitation facilities among some mission staff overseeing these activities.

- In the DRC, more than 520,000 people gained access to an improved sanitation facility in fiscal years 2012 through 2014 as a result of the mission's primary WASH activity, according to activity annual reports.[51] However, according to mission officials, the activity's reported beneficiaries included those who gained access to household-built latrines, including basic pit latrines. USAID officials in Washington, D.C., stated that community-led total sanitation, which reflects the DRC mission's approach, involves changing people's behavior regarding use of sanitation facilities by encouraging them to build latrines; however, the officials said that these latrines generally do not meet USAID's definition of improved sanitation facilities.

These officials noted that the results of community-led total sanitation efforts should be reported for USAID's WASH indicator for a community becoming open-defecation free rather than for the WASH indicator for increased access to an improved sanitation facility. According to officials at the DRC mission, the mission reported results of community-led total sanitation efforts for the indicator for first-time access to an improved sanitation facility in part because of perceived USAID headquarters emphasis on reporting numbers of beneficiaries of WASH assistance. Mission officials noted that headquarters' emphasis on numbers of beneficiaries had, to some extent, led the mission to focus on activities to increase direct access to water and sanitation rather than on efforts to improve institutions or governance.

- In Ethiopia, according to the final report for one of the activities we reviewed, 385,909 people gained access to improved sanitation in fiscal years 2009 through 2013 as a result of this activity. However, the latrines built through this activity included pit latrines without slabs (see fig. 7 for an example), which do not meet USAID's definition of improved sanitation facilities;[52] consistent with WHO and UNICEF definitions, USAID categorizes a basic pit latrine without a slab as an unimproved sanitation facility. According to Ethiopia mission officials, the activity did not fund the construction of household latrines but instead encouraged households to build their latrines with locally available materials, in consistency with Ethiopian government policy. Furthermore, mission officials noted that the reported results for this activity were based on data from the Ethiopian government's Health Management Information System, which uses an official Ethiopian government definition of improved sanitation that is not consistent with USAID's definition.
- In Indonesia, Kenya, Tanzania, and Zambia, respectively, the reported beneficiaries of activities to increase access to improved sanitation facilities included people who gained access to shared facilities, such as school toilets. For example, the Zambia mission reported that more than 133,000 people gained access to an improved sanitation facility in fiscal years 2012 through 2014 as a result of the activity focused on increasing WASH access in schools.[53] However, according to USAID guidance, shared sanitation facilities, such as those in schools and hospitals, cannot be included in the results to track progress on the number of people gaining access to improved sanitation facilities,

because this indicator is assessed at the household level. According to officials at the Zambia mission, the data they reported included the entire population of schools where sanitation facilities were built; moreover, the ratio of population to sanitation facility exceeded the national standards.[54] Mission officials noted that they focused on schools rather than households because the mission's WASH efforts are aimed at improving the environment in schools to improve education outcomes.

Evaluations of Selected USAID WASH Activities Were Generally Sound

While we found limitations in the monitoring of several of the 16 activities we reviewed, the WASH activity evaluations that we assessed were, in general, sufficiently reliable and methodologically sound for their intended purposes.[55] Ten of the 14 evaluations in our review assessed 9 of the 16 activities that we selected to examine monitoring. As noted previously in this report, performance monitoring and evaluation are separate activities; evaluations entail collection of additional information from sources that may be different from sources of monitoring data. For example, evaluators of the Ethiopia mission's primary WASH activity selected a random sample of 6 out of 41 activity sites and obtained information through focus groups with potential beneficiaries, interviews with local government officials, and personal observation of new or rehabilitated water sources. The evaluators used monitoring data where it was available for one indicator (final target and final performance data for beneficiaries gaining access to an improved water source) but independently collected data via surveys to assess the results of hygiene and sanitation component for which the implementer established a final target but did not report on annual targets or performance data. This example also illustrates that while evaluations are distinct from monitoring, they can fill some gaps in monitoring information. (See app. III for USAID's evaluation findings related to monitoring, outcomes, and sustainability.)

We assessed 14 evaluations of USAID's WASH-related activities—2 baseline evaluations, 6 midterm evaluations, and 7 final evaluations—that were conducted in fiscal years 2012 through 2014 in the nine selected countries.[56] We assessed the methodological quality of these evaluations on the basis of established evaluation principles, including the appropriateness of the evaluation design; clarity of population selection; clarity of data collection; and adequacy of support for findings, conclusions, and recommendations.[57]

- Of the 14 evaluations, 7 had clearly supported findings, conclusions, and recommendations, while 5 had certain limitations in their support for some findings, conclusions, or recommendations. These limitations included insufficient details about data collection and unclear or inappropriate criteria for population selection, given research objectives. However, we determined that, while such limitations can lead to unsupported findings and limit the usefulness of findings, conclusions, and recommendations, the limitations in these 5 evaluations were either clearly stated or otherwise did not substantially detract from the evaluations' overall purpose or utility.
- The 2 remaining evaluations had significant design limitations that suggested a lack of appropriate support for at least one finding, conclusion, or recommendation. For example, the sampling approaches for these evaluations were problematic. Both evaluations selected locations or participants on the basis of convenience, which is a nongeneralizable sampling method. This approach was not appropriate, because the evaluations' research questions were aimed at generalizing to the entire population or establishing a baseline for the future.

Thatched door opening into a pit latrine built for an activity, funded by the U.S. Agency for International Development's (USAID) Ethiopia mission, that ended in March 2013. To assess progress for this activity, the implementer used USAID's standard indicator for number of people gaining access to improved sanitation facilities, although USAID does not define a pit latrine without a slab as an improved sanitation facility.

Source: GAO. | GAO-16-81

Figure 7. Basic Household Latrine Built for a USAID Water Supply, Sanitation, and Hygiene Activity in Ethiopia.

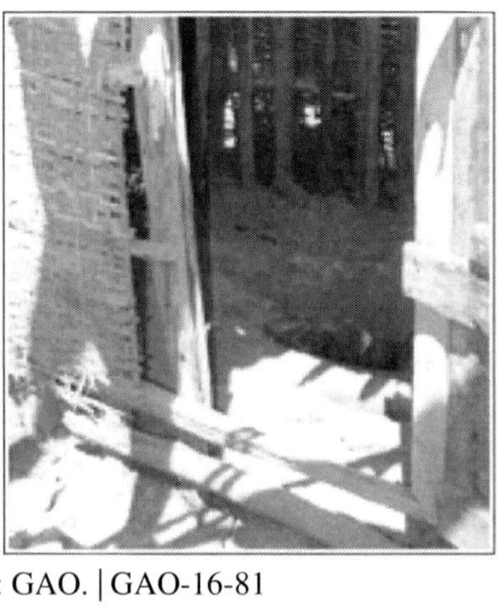

Figure 8 shows the results of our assessment of the 14 evaluations of USAID's WASH activities.

USAID mission	Evaluation title	Evaluation design appropriate	Population selection clear	Data collection clear	Findings, conclusions, and recommendations supported
Democratic Republic of the Congo	Evaluation of the Integrated Health Project	Partial	Partial	Partial	No [a]
Ethiopia	Final Performance Evaluation of the Water, Sanitation, and Hygiene Transformation for Enhanced Resilience (WATER) Project	Yes	Yes	Yes	Yes
	Terminal Evaluation Report on Your Health is in Your Hands Water, Sanitation and Hygiene program	Yes	No	No	Partial [b]
Haiti	Santé pour le Développement et la Stabilité d'Haiti (SDSH) Project Evaluation	Yes	Yes	Yes	Yes
Indonesia	Indonesia Urban Water, Sanitation, and Hygiene: Mid-term Evaluation Review	Yes	Yes	Partial	Yes
	Independent Evaluation of the Water and Sanitation Hibah Program Indonesia	Partial	No	Partial	Partial [b]
	Water, Sanitation, and Hygiene Sector Status and Trends in Indonesia	Yes	Partial	No	Partial [b]
Kenya	Mid-Term Performance Evaluation of the Sustainable Water and Sanitation in Africa (SUWASA) Project	Yes	Yes	Yes	Yes
Senegal	Mid-term Evaluation of the USAID/PEPAM Water and Sanitation Project	Yes	Partial	Partial	Partial [b]
	Yaajeende Agricultural Development Project Mid-Term Performance Evaluation	Partial	Yes	Yes	Partial [b]
Tanzania	Performance Evaluation for the Integrated Water, Sanitation and Hygiene Program: Final Report	Yes	Partial	Partial	Yes
	Tanzania Integrated Water, Sanitation and Hygiene Program: Impact Assessment Case Study	Yes	Yes	Partial	Yes
Uganda	Evaluation of WASHPlus: Supportive Environments for Healthy Communities Project	Yes	No	Partial	Yes
Zambia	End-Term Performance Evaluation for the USAID/Zambia School Water Supply and Hygiene and Quality Education	Partial	Partial	No	No [a]

Legend: Yes / No / Partial

Source: GAO analysis of U.S. Agency for International Development (USAID) information. | GAO-16-81

Source: GAO analysis of U.S. Agency for International Development (USAID) information. | GAO-16-81

Notes: When determining the overall quality of an evaluation, we considered whether design, population selection, and data gathering were sufficient to support findings, conclusions, and recommendations. If limitations in these areas were clearly explained in the evaluation, or if they did not substantially affect findings, conclusions, or recommendations, our overall judgment of evaluation quality was positive.

[a] The evaluation had a significant design limitation that suggested a lack of appropriate support for at least one finding, conclusion, or recommendation.

[b] The evaluation had certain limitations in its support for some findings, conclusions, or recommendations; however, these limitations did not substantially detract from the overall purpose or utility of the evaluation.

Figure 8. Assessment of Evaluations of USAID's Water Supply, Sanitation, and Hygiene (WASH) Activities in Fiscal Years 2012-2014.

We found that in some cases, USAID incorporated evaluation results to improve activity monitoring. Our interviews with mission officials and reviews of monitoring documents indicated that the selected USAID missions modified ongoing or planned activities—for at least 6 of the 10 we reviewed that had relevant evaluations—on the basis of evaluation results.[58] For example, the midterm evaluation of the primary WASH activity in Indonesia recommended reducing the activity's target of 40,000 households willing to pay for sanitation improvements, because of the challenges involved in meeting this goal. As a result, the Indonesia mission modified its agreement with the implementer and reduced the target to 15,000 households. Mission officials noted that the midterm evaluation also provided specific recommendations for future USAID projects that would be taken into consideration during the design process for follow-on WASH investments in Indonesia.

Of the 7 final performance evaluations of WASH activities that we reviewed, 3 were conducted before the activity's completion, 3 were conducted within 1 month of the activity's completion, and 1 was conducted within 3 months of the activity's completion. As a result, these studies were not set up to allow for an assessment or the longer-term sustainability of these projects. To enable assessments of activities' impact and sustainability, USAID has identified plans to conduct evaluations for some WASH activities several years after project completion. Specifically, USAID's Water Strategy notes that the agency plans to conduct assessments of WASH sustainability beyond the typical USAID project cycle and to provide support for issues that arise subsequent to completion of WASH activities. Additionally, in an April 2014 document, USAID noted plans to increase its focus on sustaining development outcomes by, among other things, conducting evaluations 3 to 5 years after project conclusion.[59] According to USAID, such long-term evaluations provide opportunities to explore the impact of interventions and may contribute to a deeper understanding of programmatic risk.

CONCLUSION

Since 2005, USAID has reported that millions have gained access to improved drinking water and sanitation facilities as a result of its assistance. In 2013, USAID issued its first Water Strategy, articulating goals for the provision of sustainable WASH assistance. In response, USAID missions in a number of priority countries are developing a more strategic approach to

WASH, and USAID's Office of Water and missions have begun taking steps to address sustainability of WASH investments. Nevertheless, these efforts are still in the early stages.

However, limitations in some missions' monitoring and reporting for WASH activities that we reviewed call into question USAID's ability to reliably assess and report progress toward its strategic WASH goals. Unless USAID identifies and addresses factors contributing to missions' inconsistent adherence to guidance regarding establishing annual targets for key WASH indicators for all WASH activities and disaggregating activity data by gender, USAID cannot reliably measure these activities' contributions to achieving WASH goals or toward gender equality and women's empowerment. Moreover, unless USAID identifies and addresses factors contributing to missions' inconsistent adherence to guidance for verifying beneficiaries of water activities and accurately reporting beneficiaries of sanitation activities, USAID cannot ensure the accuracy of its annual reports regarding progress in increasing access to safe water and sanitation.

Recommendations for Executive Action

To effectively address limitations in missions' monitoring and reporting of USAID's WASH activities, we are making the following two recommendations to the USAID Administrator. Specifically, with respect to inconsistent adherence to agency guidance for establishing annual targets, for reporting gender disaggregated data, for verifying beneficiaries of water activities, and for accurately reporting beneficiaries of sanitation activities, USAID should

- identify factors contributing to missions' inconsistent adherence to agency guidance and
- take steps to address these factors.

Agency Comments and Our Evaluation

We provided a draft of this report to State and USAID. USAID provided written comments as well as technical comments that we incorporated as appropriate. State did not provide comments.

In its written comments, USAID generally concurred with our recommendations and outlined steps it is taking to address our second recommendation. Following are highlights of USAID's comments, with our evaluation:

1. USAID noted that our definition of "obligations" excludes bilateral obligations that have not yet been sub-obligated at the mission-level for WASH activities. As our report states, for the purposes of our review, we defined obligations as orders placed, contracts awarded, and similar transactions during a given period that will require payments during the same or a future period. USAID categorizes these as "sub-obligations," because it considers these funds to have been obligated through a bilateral agreement with the host country.
   In reporting funding for nine selected missions' WASH activities, we generally reported total allocations for WASH activities ongoing at each mission in fiscal years 2012 through 2014, including, when applicable, funding for years before fiscal year 2012 and through fiscal year 2014. (State and USAID define allocations as the distribution of resources to bureaus and operating units by foreign assistance account.) We included obligations when USAID data did not show WASH activity allocations.
2. USAID stated that it develops targets through multiple processes and that it had provided us with performance plans and reports, which include future-year targets for missions. On the basis of USAID's technical comments on our draft report, we revised our assessment to acknowledge that the mission-wide performance management plans and reports for fiscal years 2012 and 2013 included annual targets for the two activities we had selected for Zambia ("Schools Promoting Learning Achievement through Sanitation and Hygiene" and "Partnership for Integrated Social Marketing"). Although we included this information in response to USAID's comments, it is important to note that neither the monitoring plans nor the annual reports for these two activities included annual targets. According to USAID policy (ADS ch. 203), activity-level monitoring plans feed into project-level monitoring plans and mission-wide performance management plans. Therefore, the absence of annual targets in activity-level monitoring plans calls into question the completeness of the Zambia mission-wide performance management plan. Additionally, given that mission-wide performance plans and reports provide only aggregate target and

performance data, it may not be possible to identify activity- specific targets or performance data from these reports.

3. USAID stated that our report does not reference certain key documents that USAID had provided. Although USAID's letter does not specify which key documents it is referring to, in its technical comments, USAID refers to mission-wide performance plans and reports and to State's annual Water Key Issue Definition guidance. In our report, we refer to performance plans and reports as mission-level annual reporting. In appendix I of our report, we also note that to assess data reliability, we compared implementer reporting on an activity with mission-level annual reporting for the nine selected USAID missions. In addition, our report notes that we reviewed State's annual Water Key Issue Definition guidance and assessed the extent to which WASH activities in the nine missions generally adhered to this guidance's requirement regarding activities for which funds can be attributed to the congressional spending requirement for water and sanitation. We described the content of State's guidance in the background section of our report, where we added a footnote, in response to USAID comments, to more clearly identify the guidance documents we referred to.
4. USAID stated that the inconsistencies identified in our draft report pertaining to the inaccurate categorization of results were the byproduct of isolated incidences of reporting against incorrect indicators. We found inaccuracies in six of eight activities we reviewed that reported on the standard indicator "number of people gaining access to an improved sanitation facility." While USAID's comments confirmed these inaccuracies, the agency has not provided any documentation or other support for its statement that they represent isolated incidences.
5. USAID stated that missions are allowed to use custom indicators to track results against water-directive funded activities. We did not intend to imply that use of custom indicators was not in compliance with State and USAID requirements. While we assessed the selected missions' use of a number of WASH indicators to track progress for 16 selected activities, our review focused in particular on the standard indicators for access to improved water source and improved sanitation. As our report notes, State and USAID have reported overall progress on WASH using these two indicators, and USAID's *Water and Development Strategy (2013-2018)* uses these two

indicators to project numbers of beneficiaries during the strategy's 5-year period (i.e., at least 10 million persons would receive sustainable access to improved water supply and 6 million persons would receive sustainable access to improved sanitation). Nevertheless, in our report's background section, we include a table of standard and custom indicators that USAID allows missions to use in reporting on results of WASH activities. Additionally, on the basis of a follow-up discussion with USAID, we understand that the agency was particularly concerned that our report might seem to imply that its Indonesia mission's use of a custom indicator for verifying households for the mission's Water Hibah activity was not in compliance with agency guidance. As a result, we have added a note to figure 6 to clarify that USAID guidance allows missions the flexibility to report using custom indicators.

6. USAID noted that it informed us that verification information could be found in data quality assessments and site visit reports from missions such as Senegal. During our audit work, after the mission informed us that it had conducted verification of beneficiaries through these assessments, we requested data quality assessments from the USAID mission in Senegal. However, the mission did not provide these assessments to us. In the absence of documentation of verification of beneficiaries, we maintain that it is unclear whether the mission verified the number of beneficiaries gaining access to an improved water source.

David Gootnick, Director
International Affairs and Trade

APPENDIX I: OBJECTIVES, SCOPE, AND METHODOLOGY

In this report, we (1) describe the types of water supply, sanitation, and hygiene (WASH) activities the U.S. Agency for International Development (USAID) has implemented in selected countries and the funding it has provided for these activities; (2) assess the extent to which USAID guidance has informed the agency's efforts to plan and implement WASH activities in these countries; and (3) assess USAID's monitoring and evaluation of selected WASH activities.[60]

We focused our review on WASH activities that USAID missions had implemented in 9 selected countries: the Democratic Republic of the Congo (DRC), Ethiopia, Haiti, Indonesia, Kenya, Tanzania, Senegal, Uganda, and Zambia.[61] We selected these countries from the list of 22 countries that USAID designated as priority countries for WASH assistance in fiscal year 2014.[62] We based our country selection primarily on the amounts of funding that the missions in these countries attributed to the congressional spending requirement for international water and sanitation assistance in fiscal years 2012 through 2013.[63] During that period, WASH assistance in the 9 selected countries accounted for 53 percent—$155 million—of the funding attributed to the spending requirement by the missions in all tier 1 and tier 2 priority countries.[64] To address our three reporting objectives, we interviewed USAID and Department of State (State) officials and conducted semi-structured telephone interviews with officials at the nine USAID missions regarding their efforts to plan, monitor, and evaluate WASH activities. Because we judgmentally selected the nine USAID missions for our review, our findings from these interviews cannot be generalized to all USAID missions. In addition, we conducted fieldwork in Tanzania and Ethiopia from January 26, 2015, to February 6, 2015. We selected these countries for fieldwork based on factors such as level of funding and type of WASH activities implemented in each country. For example, the Ethiopia mission attributed the most funding to the congressional spending requirement in fiscal years 2012 through 2013 for a range of WASH activities, and the Tanzania mission has implemented one long-standing (since 2009) primary activity focused on WASH.

To describe the types of WASH activities that USAID has implemented and the funding it has provided for these activities, we developed a data collection instrument to obtain funding data and descriptive information from each mission regarding its WASH activities that were ongoing or planned during fiscal years 2012 through 2014. Each mission's total funding for WASH activities generally represents allocations for WASH components of activities that were ongoing at the mission at any time in fiscal years 2012 through 2014, including, when applicable, WASH funding for years before fiscal year 2012 and through fiscal year 2014. We included obligations in cases for which USAID was not able to provide allocations for WASH activities, such as when a USAID mission obligated unplanned funding to an activity for WASH.[65] To assess the reliability of the data and information we obtained, we reviewed documentation and interviewed agency officials to identify and correct any missing data and any errors. We determined that the data and information we gathered were sufficiently reliable to provide general information about the types of activities implemented and approximate funding provided for these activities.

To assess the extent to which USAID guidance informed planning and implementation of WASH activities, we reviewed USAID guidance, mission-level WASH plans, and other documents. We reviewed USAID guidance including: (1) USAID's Automated Directives System (ADS), chapter 201, which contains agency policies and procedures and includes guidance on strategic planning and project and activity design; (2) USAID's *Water and Development Strategy, 2013-2018* (Water Strategy); and (3) USAID's *Water and Development Strategy Implementation Field Guide* (Field Guide). We assessed the status of the USAID missions' efforts to develop WASH plans, steps that they have taken to implement these plans, and steps that they have taken to adopt the Water Strategy's principles and approach for their recent or planned WASH activities. Missions' WASH plans may consist of one or more project appraisal documents (PAD), which document a project's design and expected results, as well as a WASH sector assessment or other related documents that describe the mission's strategic approach to WASH. To assess steps that missions have taken to address sustainability, we reviewed documents that included PADs and sustainability assessments; we also reviewed award documents for selected activities.

We generally selected two activities per country on the basis of factors such as the level of funding allocated and types of WASH activities implemented in fiscal years 2012 through 2014. We based our activity selection on funding data reported in State's Foreign Assistance Coordination and Tracking System for fiscal years 2012 and 2013, data included in missions' operational plans for fiscal year 2014, and discussions with mission officials about their WASH activities between fiscal years 2012 through 2014. The activities we selected included those to which the missions in the selected countries had allocated the largest amounts of WASH funding, with one exception: In Uganda, the selected activities did not include the Uganda mission's activity that received the largest allocations of WASH funding in fiscal years 2012 through 2014, because the mission had allocated the majority of funding for its largest WASH activity before fiscal year 2012, did not initially inform us of this activity, and did not provide the prior-year funding data until after we had made our activity selection. We selected one activity in Tanzania, because the Tanzania mission implemented only one primary

WASH activity during the period we reviewed. We initially selected three WASH activities in Haiti.[66] However, the Haiti mission subsequently informed us that two of the selected activities did not have a WASH component and that the mission had incorrectly attributed WASH funds to these activities and planned to take corrective action to ensure use of attributed funds for WASH. As a result, we reviewed one WASH activity for Haiti.

To assess monitoring of WASH activities in the nine countries, we obtained and analyzed documentation for selected WASH activities, including award agreements and modifications, performance management plans, monitoring and evaluation plans, quarterly and annual monitoring reports, and annual funding data. We compared monitoring plans with quarterly and annual reports to determine whether specific WASH indicators had been identified for the activity. Since several activities had multiple WASH indicators, we focused our analysis on standard indicators related to water, sanitation, or hygiene that were used to monitor the activity. When an activity's monitoring plans included no standard WASH indicators, we identified at least one activity indicator related to access to water, access to sanitation, or hygiene improvement, as relevant. For the purposes of this report, we refer to the indicators that we identified for our review as key WASH indicators. We reviewed each activity's monitoring plan and monitoring reports to determine whether (1) annual targets were identified for the key WASH indicators and (2) results or performance data for each key WASH indicator were reported on an annual basis, if applicable. Further, we reviewed monitoring reports to assess whether they included gender-disaggregated data for indicators on number of people gaining access to an improved drinking water source or number of people gaining access to an improved sanitation facility. We compared the reported performance data for three key WASH indicators—number of people gaining access to improved drinking water source, number of people gaining access to an improved sanitation facility, and number of liters of drinking water disinfected with point-of-use treatment products as a result of U.S. government assistance—with USAID and State's guidance for these indicators to assess the extent to which the reported data conformed to the definitions in the guidance document. Although USAID and State guidance required verification of estimated beneficiaries for the indicator for access to an improved water source, the guidance did not require verification of estimated numbers of liters of water purified. In addition to assessing activity performance data against agency guidance, we conducted internal consistency checks to assess the reliability of reported data. For example, to the extent feasible, we compared implementer reporting on an activity with mission-level annual reporting and USAID-wide annual reporting on WASH indicators. Because we found data inconsistencies for several activities, as noted in the report, we did not use the performance data to report the extent to which activities met intended targets.

To assess USAID evaluations for WASH activities in the nine countries, we selected all evaluations for WASH activities completed in fiscal years 2012 through 2014. We identified 14 completed evaluations, including evaluations for 10 of the 16 activities we selected to assess monitoring. To assess the soundness of the evaluations, we reviewed background information about programs and evaluation questions, assessed evaluation design and process, and considered the evaluation results and limitations of each study. Two GAO specialists conducted these assessments independently, using a tool that incorporated key elements of USAID's scope-of-work checklist for evaluations and considered various aspects of these issues. The two specialists compared the results of their independent assessments and came to agreement about all conclusions. We did not assess selected USAID missions' compliance with the requirement to conduct evaluations.[67]

- To review background information about the evaluated programs, we considered whether the evaluations considered evaluator independence, program objectives and mechanisms, and evaluation goals. We also assessed whether the relationship between the evaluation objectives and program design was clear and appropriate.
- To assess evaluation design and process, we considered whether the evaluations clearly described their design; whether appropriate methods were used to select the locations and people covered by the study, including whether the evaluations provided sufficient detail about sampling methods; whether measures used were clearly related to evaluation questions; and whether data collection and analysis were sufficient and appropriate. We also assessed whether selection methods, sample sizes, criteria, measures, data collection, analysis, and overall design were appropriate, given the evaluation objectives.
- To consider evaluation results and limitations, we determined whether findings, conclusions, recommendations, and lessons learned were clearly stated, whether stakeholders were given an opportunity to comment on the results, and whether evaluations provided information about how results should be used. We also assessed whether the evaluations clearly and sufficiently described assumptions and limitations of their design and results, including potential biases,

confounding variables, unintended consequences, alternative explanations, and methodological limitations. We assessed whether any findings, conclusions, recommendations, or lessons learned were appropriately supported and caveated, given the evaluation design. To summarize evaluation results, we determined whether the sections of evaluations related to program monitoring, outcomes, or sustainability contained descriptive information or appropriately supported findings, conclusions, recommendations, or lessons learned.

We conducted this performance audit from July 2014 to October 2015 in accordance with generally accepted government auditing standards. Those standards require that we plan and perform the audit to obtain sufficient, appropriate evidence to provide a reasonable basis for our findings and conclusions based on our audit objectives. We believe that the evidence obtained provides a reasonable basis for our findings and conclusions based on our audit objectives.

APPENDIX II: USAID WASH ACTIVITIES IN NINE SELECTED COUNTRIES

Following are funding data and descriptive information about water, sanitation, and hygiene (WASH) activities implemented in fiscal years 2012 through 2014 in the nine countries that we selected for our review— the Democratic Republic of the Congo, Ethiopia, Haiti, Indonesia, Kenya, Tanzania, Senegal, Uganda, and Zambia.

Democratic Republic of the Congo

The U.S. Agency for International Development (USAID) mission in the Democratic Republic of the Congo implemented seven WASH activities in fiscal years 2012 to 2014, with WASH activity funding totaling $26,125,075 (see table 4). The mission reported two primary WASH activities—the Integrated Health Project and Sustainable WASH Interventions: Healthy Villages Program in Two Health Zones—focused on improving access to water, sanitation, and hygiene services in target locations. Allocations of WASH funding for these two activities totaled $22,518,470.

Table 4. USAID WASH Activities Implemented in the Democratic Republic of the Congo, Fiscal Years 2012-2014

Activity (duration)	Description	Activity focus	Geographic focus[a]	Total funding[b]
Integrated Health Project (2010-2015) c	Activity works at the provincial and community levels to improve access to health services, including water, sanitation, and hygiene. For example, the activity has supported water points, latrines, and WASH behavior- change communication.	BCC, CB, IC	Rural, peri-urban	**$13,377,369**
Sustainable WASH Interventions: Healthy Villages Program in Two Health Zones, Provinces of Katanga and Kasai Oriental (2010-2015)c	Activity supports access to improved water and sanitation in selected villages. Efforts include capacity building to provide WASH technical skills and project management skills to local technicians and community workers, support for water points and latrine construction, and support for improved national-level coordination on WASH issues.	BCC, CB, IC, PG, TA	Rural, peri-urban	**9,141,101**
Advancing Social Marketing for Health (2009-2013)	Activity promoted and distributed point-of-use water purification products and conducted behavior-change communication for improved WASH practices.	BCC, CB	Peri-urban, urban	**1,400,000**
Asili Social Enterprise Zones (2014-2018)	Activity supports increased access to, and use of, clean water points.	BCC, F, IC	Rural, peri-urban	**1,000,000**
Communication for Change (2012-2015)	Activity supports behavior-change communication to improve health outcomes related to WASH. For example, activity produces WASH educational materials for use by USAID partners and other stakeholders nationwide.	BCC, CB, PG	Rural, peri-urban, urban	**726,605**

Activity (duration)	Description	Activity focus	Geographic focus[a]	Total funding[b]
Health Systems 20/20 (2010-2012)	Activity provided support for national level WASH coordination.	PG	Rural, peri-urban, urban	**300,000**
Measure Demographic and Health Surveys III ICF MACRO (2012-2015)	Activity supports national survey on household access to water and sanitation and WASH behaviors, to inform programming and measure progress related to WASH.	CB	Rural, peri-urban, urban	**180,000**
Total activities 7				**Total funding $26,125,075**

Legend:

BCC = behavior-change communication

CB = capacity building

F = financing

IC = infrastructure construction

PG = policy and governance

TA = technical assistance

USAID = U.S. Agency for International Development

WASH = water supply, sanitation, and hygiene

Source: GAO analysis of USAID documents and data. | GAO-16-81

[a] According to a USAID official, USAID missions do not employ a universal definition for rural, peri-urban, or urban, and missions are encouraged to defer to host-country census authorities' definitions of these terms.

[b] Funding shown generally represents allocations for WASH components of activities and includes, when applicable, WASH funding for years before fiscal year 2012 and through fiscal year 2014. We show obligations for activities for which USAID data do not show WASH allocations, such as activities for which a USAID mission had obligated unplanned funding for WASH.

[c] We selected this activity for our assessment of USAID's planning, monitoring, and—where applicable—evaluation of WASH.

Ethiopia

The USAID mission in Ethiopia implemented 11 WASH activities in fiscal years 2012 through 2014, with WASH activity funding totaling $24,055,770 (see table 5). The mission's largest WASH activity, called Water, Sanitation, and Hygiene Transformation for Enhanced Resilience, included $10,984,723 in funding for WASH and focused on providing water infrastructure in pastoralist areas.

Table 5. USAID WASH Activities Implemented in Ethiopia, Fiscal Years 2012-2014

Activity (duration)	Description	Activity focus	Geographic focus[a]	Total funding[b]
Water, Sanitation, and Hygiene Transformation for Enhanced Resilience (2011-2015)[c]	Activity supports integrated WASH services in the Ethiopian drylands, including new and rehabilitated water systems and extension services.	BCC, CB, IC, TA	Rural	**$10,984,723**
Your Health is in Your Hands Water, Sanitation and Hygiene program (2009-2013)[c]	Activity supported efforts to improve the use of safe water, sanitation, and hygiene services and practices at households, schools, and health facilities.	BCC, CB, IC, TA	Rural, peri-urban	**4,300,000**
Empowering New Generations to Improve Nutrition and Economic Opportunities (2011-2016)	As the mission's main Feed the Future nutrition activity, this activity is working to integrate WASH through subsidizing water filters for home use and by conducting sanitation marketing.	BCC, F, PG	Rural	**3,400,000**
American Councils on Education (2010-2015)	Activity aims to build local capacity in the areas of water, public health, sanitation, and resource management.	CB	Rural, urban	**1,500,000**
Transforming Education for Adults and Children in the Hinterlands II (2009-2015)	Activity provides capacity building related to water, sanitation and health for alternative basic education centers.	CB, BCC	Rural, peri-urban	**1,100,000**
Global Team for Local Initiatives (2009-2012)	Activity included construction of hand dug wells and community training related to hygiene behaviors, including hand and face washing. Activity also included construction of pit latrines to protect water access points.	BCC, IC	Rural	**586,966**
Oromo Self-Help Organization (2013-2016)	WASH implemented as component of broader activity focused on mitigating fluorosis, a disease caused by consumption of excess fluoride.	BCC, CB, IC	Peri-urban, urban	**549,982**
Strengthening Institutions for	Activity provides support to government and local	IC	Rural	**500,000**

Activity (duration)	Description	Activity focus	Geographic focus[a]	Total funding[b]
Peace and Development (2004-2012)	institutions to build capacity for conflict mitigation, including conflict related to potable water.			
Higher Education for Development (2010-2015)	Activity provides training to students and support to augment curricula at Ethiopian universities, particularly curricula related to water and wastewater treatment and water and public health. Activity also supported inauguration of the Ethiopian Institute of Water Resources.	CB, TA	Peri-urban, urban	**496,355**
Research-inspired Policy and Practice Learning in Ethiopia (2013-2015)	Activity includes conducting survey of community and institutions on hygiene and sanitation waste disposal, establishment of environmental sanitation learning and practice alliances, and training for community groups on productive reuse of solid waste and waste water.	BCC, CB	Urban	**337,744**
Tsehai Loves Learning (2013-2016)	Activity produces an award-winning, Ethiopia- specific children's television show to promote early literacy skills and health messages, including WASH segments related to latrine use, hand and face washing, and food safety.	BCC	Peri-urban, urban	**300,000**
Total activities 11				**Total funding $24,055,770**

Legend:
BCC = behavior-change communication
CB = capacity building
F = financing
IC = infrastructure construction PG = policy and governance TA = technical assistance
USAID = U.S. Agency for International Development
WASH = water supply, sanitation, and hygiene
Source: GAO analysis of USAID documents and data. | GAO-16-81

[a] According to a USAID official, missions do not employ a universal definition for rural, peri-urban, or urban, and missions are encouraged to defer to host-country census authorities' definitions of these terms.

[b] Funding shown generally represents allocations for WASH components of activities and includes, when applicable, WASH funding for years before fiscal year 2012 and through fiscal year 2014. We included obligations for activities for which USAID data do not show WASH allocations, such as activities for which a USAID mission had obligated unplanned funding for WASH.

[c] We selected this activity for our assessment of USAID's planning, monitoring, and—where applicable—evaluation of WASH.

Source: GAO. | GAO-16-81

Figure 9. Drinking Water Tank and Taps Constructed as Part of USAID's WASH Activity in Buda Megada Village, Bule Hora District, Oromiya, Ethiopia.

Figure 9 shows a drinking water reservoir and a water point constructed through the Ethiopia mission's largest WASH activity.

Haiti

The USAID mission in Haiti implemented six WASH activities in fiscal years 2012 through 2014, with WASH activity funding totaling $4,436,481 (see table 6). The mission's largest WASH activity, called Santé pour le Développement et la Stabilité d'Haiti, included $1,711,000 in funding for the WASH component of a broader activity that focused on improving the health status of Haitians through improved primary care, referral networks, and management practices at health facilities and in communities.

Table 6. USAID WASH Activities Implemented in Haiti, Fiscal Years 2012-2014

Activity (duration)	Description	Activity focus	Geographic focus[a]	Total funding[b]
Santé pour le Développement et la Stabilité d'Haiti (2007-2012)	Activity supported improved health status of Haitians through improved primary care, referral networks, and management practices, including WASH, at health facilities and in communities.	CB, TA	Rural, peri-urban, urban	**$1,711,000**
Services de Santé de Qualité pour	Activity supports improved health status of Haitians	CB, TA	Rural, peri-urban, urban	**1,324,487**

Activity (duration)	Description	Activity focus	Geographic focus[a]	Total funding[b]
Haiti (2013-2016)[c]	through improved primary care, referral networks, and management practices, including WASH, at health facilities and in communities.			
PROMARK II and AIDSTAR (2009-2014)	Activity supported healthy family planning and safe water systems behavior to improve health outcomes.	CB	Rural, peri-urban, urban	**512,384**
Community Health and AIDS Mitigation Program (2009-2013)	Activity supported increased access to and use of basic health services, including WASH, through community-based care and support services for vulnerable populations.	CB	Rural, peri-urban, urban	**488,610**
Health Information Systems (2013-2017)	Activity strengthens capacity of the Haitian government to manage program resources and monitor patient outcomes related to cholera. Co-managed by the Centers for Disease Control and Prevention.	CB	Rural, peri-urban, urban	**200,000**
Health Through Walls Prison Project (2011-2015)	Activity provides comprehensive care to prisoners in Haiti's National Penitentiary, women's and children's prisons, and included a WASH component to prevent spread of cholera.	IC	Rural, peri-urban, urban	**200,000**
Total activities 6				**Total funding $4,436,481**

Legend:

BCC = behavior-change communication

CB = capacity building

F = financing

IC = infrastructure construction PG = policy and governance TA = technical assistance

USAID = U.S. Agency for International Development

WASH = water supply, sanitation, and hygiene

Source: GAO analysis of USAID documents and data. | GAO-16-81

[a] According to a USAID official, USAID missions do not employ a universal definition for rural, peri- urban, or urban, and missions are encouraged to defer to host-country census authorities' definitions of these terms.

[b] Funding shown generally represents allocations for WASH components of activities and includes, when applicable, WASH funding for years before fiscal year 2012 and through fiscal year 2014. We included obligations for activities for which USAID did not show WASH allocations, such as activities for which a USAID mission had obligated unplanned funding for WASH.

[c] We selected this activity for our assessment of USAID's planning, monitoring, and—where applicable—evaluation of WASH.

Indonesia

The USAID mission in Indonesia implemented five WASH activities in fiscal years 2012 through 2014, with WASH activity funding totaling $53,401,700. The mission's largest WASH activity, called Indonesia Urban Water, Sanitation, and Hygiene, included $38,696,403 in WASH funding and focused on providing access to water and sanitation facilities in urban areas.

Table 7. USAID WASH Activities Implemented in Indonesia, Fiscal Years 2012-2014

Activity (duration)	Description	Activity focus	Geographic focus[a]	Total funding[b]
Indonesia Urban Water, Sanitation, and Hygiene (2011-2016)[c]	Activity provided support for national and local governments and utilities to mobilize demand for WASH, increase service delivery, and provide policy and financing options for WASH in urban areas.	BCC, CB, F, PG, TA	Urban	**$38,696,403**
Urban Household Water Supply in Indonesia (Water Hibah) (2011-2015)[c]	Activity supported water connections in urban areas.	IC	Urban	**10,000,000**
Development Grant Program High Five (2011-2014)	Activity implemented community-led total sanitation in urban areas.	BCC, TA	Urban	**2,000,000**
Development Grant Program IWINS (2013-2016)	Activity aims to increase the urban poor's access to a safe water supply and improved sanitation facilities.	CB, TA	Urban	**1,405,297**
Development Grant Program Water SMS (2010-2013)	Activity developed a system to enable feedback on services for water utilities.	CB	Urban	**1,300,000**
Total activities 5				**Total funding $53,401,700**

Legend:
BCC = behavior-change communication
CB = capacity building
F = financing
IC = infrastructure construction PG = policy and governance TA = technical assistance
USAID = U.S. Agency for International Development
WASH = water supply, sanitation, and hygiene
Source: GAO analysis of USAID documents and data. | GAO-16-81

[a] According to a USAID official, USAID missions do not employ a universal definition for rural, peri- urban, or urban, and missions are encouraged to defer to host-country census authorities' definitions of these terms.

[b] Funding shown generally represents allocations for WASH components of activities, and includes, when applicable, WASH funding for years before fiscal year 2012 and through fiscal year 2014. We included obligations for

activities for which USAID data did not show WASH allocations, such as activities for which a USAID mission had obligated unplanned funding for WASH.

[c] We selected this activity for our assessment of USAID's planning, monitoring, and—where applicable—evaluation of WASH.

Kenya

The USAID mission in Kenya implemented 19 WASH activities in fiscal years 2012 through 2014, with WASH activity funding totaling $18,245,655. The mission's largest WASH activity, called AIDS, Population and Health Integrated Assistance Program (APHIA) Plus Northern Arid Lands Service Delivery, included $3,024,887 in funding for WASH. The activity focuses on HIV/AIDS, maternal and child health, WASH, and nutrition for orphaned and vulnerable children. The mission reported five similar activities targeting other regions of the country.

Table 8. USAID WASH Activities Implemented in Kenya, Fiscal Years 2012-2014

Activity (duration)	Description	Activity focus	Geographic focus[a]	Total funding[b]
AIDS, Population and Health Integrated Assistance Program (APHIA) Plus Northern Arid Lands Service Delivery (2012-2017)[c]	Activity addresses HIV/AIDs, maternal and child health, WASH and nutrition in health facilities, schools, and at the community and household level.	BCC, CB, IC, PG, TA	Rural, peri-urban, urban	**$3,024,887**
APHIA Plus Nyanza/Western (2011-2015)	Activity addresses HIV/AIDs, maternal and child health, WASH and nutrition in health facilities, schools, and at the community and household level.	BCC, CB, IC, TA	Rural, peri-urban, urban	**2,562,254**
Kenya Agriculture Value Chain Enterprises (2013-2018)	The WASH component in this Feed The Future activity is to provide drinking water for the communities	BCC, CB, F, IC, PG, TA	Rural, peri-urban, urban	**2,000,000**

Table 8. (Continued)

Activity (duration)	Description	Activity focus	Geographic focus[a]	Total funding[b]
Kenya Agriculture Value Chain Enterprises (2013-2018)	The WASH component in this Feed The Future activity is to provide drinking water for the communities	BCC, CB, F, IC, PG, TA	Rural, peri-urban, urban	**2,000,000**
Agile and Harmonized Assistance to Devolved Institutions (2014-2018)	Activity primarily focuses on governance and includes support for the national policy and legal framework for WASH, as well as capacity building for selected communities related to water laws, policies, regulations and strategies.	BCC, CB, IC, PG, TA	Rural, peri-urban, urban	**1,500,000**
Kenya Arid Lands Disaster Risk Reduction (2012-2015)	Activity works to increase access to clean water and improve sanitation and hygiene in Kenya's arid lands.	BCC, CB, F, IC, PG, TA	Rural, peri-urban, urban	**1,500,000**
APHIA Plus Eastern/Central/ Kamili (2011-2015)	Activity addresses HIV/AIDs, maternal and child health, WASH and nutrition in health facilities, schools, and at the community and household level.	BCC, CB, IC, TA	Rural, peri-urban, urban	**1,188,906**
APHIA Plus Rift Valley (2011-2015)	Activity addresses HIV/AIDs, maternal and child health, WASH and nutrition in health facilities, schools, and at the community and household level.	BCC, CB, IC, TA	Rural, peri-urban	**1,000,843**
APHIA Plus Nairobi/Coast (2011-2013)	Activity addresses HIV/AIDs, maternal and child health, WASH and nutrition in health facilities, schools, and at the community and household level.	BCC, CB, IC, TA	Rural, peri-urban, urban	**684,304**
Maternal and Child Health Integrated Program (2008-2014)	Activity addressed maternal and child health, nutrition, and WASH in health facilities and communities.	BCC, CB, IC, TA	Rural, peri-urban	**670,000**
Water, Sanitation, and Hygiene Plus (2011-2016)	Activity focuses on providing technical assistance and building the capacity of the ministry of health for WASH.	BCC, CB, IC, PG, TA	Rural, peri-urban, urban	**620,000**
Financial Inclusion for Rural Microenterprise	Activity motivates financial institutions to finance the	BCC, CB, F,	Rural, peri-	**600,000**

Activity (duration)	Description	Activity focus	Geographic focus[a]	Total funding[b]
(2012-2017)	WASH sector, especially to increase water access.	IC, PG, TA	urban, urban	
Maternal and Child Health Integrated Program/Jhpiego (2014-2019)	Activity addresses maternal and child health, nutrition, and WASH at health facilities, schools, and in communities.	BCC, CB, IC, TA	Rural, peri-urban, urban	**600,000**
Laikipia Natural Resource Management and Biodiversity Conservation Program (2009-2015)	Activity includes a small WASH component related to drinking water supply. Activity also addresses water resource management for downstream biodiversity.	BCC, CB, F, IC, PG, TA	Rural, peri-urban, urban	**500,000**
Sustainable Water and Sanitation in Africa Project to Support Urban WASH in Kenya (Kenya 2) (2013-2017)[c]	Activity supports long-term financial sustainability for water and sanitation service delivery, working closely with utilities.	BCC, CB, F, IC, PG, TA	Rural, peri-urban, urban	**500,000**
National Training, Funzo Kenya Project (2012-2017)	Activity supports training of health service providers in various technical areas, including WASH.	BCC, CB, TA	Rural, peri-urban, urban	**400,000**
Development Credit Authority Program (2014-2017)	Activity supports private sector financing for WASH through credit guarantees.	BCC, CB, F, IC, PG, TA	Urban	**300,000**
APHIA Plus Nairobi Coast Follow-on (2011-2013)	Activity supported maternal child health, HIV/AIDs, nutrition, and WASH activities in health facilities and communities.	BCC, CB, IC, TA	Rural, peri-urban, urban	**294,461**
University of North Carolina/Pima/Measure Evaluation Project (2012-2017)	Activity provides technical support to the government of Kenya and implementing partners for data collection systems for decision making related to WASH.	BCC, CB, IC, PG, TA	Rural, peri-urban, urban	**200,000**
Academic Model Providing Access to Healthcare Plus (2012-2017)	Activity addresses HIV/AIDs, maternal and child health, WASH and nutrition in health facilities, schools, and at the community and household level.	BCC, CB, IC, TA	Rural, peri-urban, urban	**100,000**
Total activities 19				**Total funding $18,245,655**

Legend:

BCC = behavior-change communication

CB = capacity building

F = financing

IC = infrastructure construction PG = policy and governance TA = technical assistance
USAID = U.S. Agency for International Development
WASH = water supply, sanitation, and hygiene
Source: GAO analysis of USAID documents and data. | GAO-16-81
[a] According to a USAID official, USAID missions do not employ a universal definition for rural, peri- urban, or urban, and missions are encouraged to defer to host-country census authorities' definitions of these terms.
[b] Funding shown generally represents allocations for WASH components of activities, and includes, when applicable, WASH funding for years before fiscal year 2012 and through fiscal year 2014. We included obligations for activities for which USAID data do not show WASH allocations, such as activities for which a USAID mission had obligated unplanned funding for WASH.
[c] We selected this activity for our assessment of USAID's planning, monitoring, and—where applicable—evaluation of WASH.

Senegal

The USAID mission in Senegal implemented five WASH activities in fiscal years 2012 through 2014, with WASH activity funding totaling $27,616,000. The mission's largest WASH activity, called Senegal Millennium Water and Sanitation Program, included $20,866,000 in funding for WASH. The activity focused on governance and management, the creation of local business opportunities, increasing demand for clean water and sanitation, the construction of infrastructure, and hygiene.

Table 9. USAID WASH Activities Implemented in Senegal, Fiscal Years 2012-2014

Activity (duration)	**Description**	**Activity focus**	**Geographic focus[a]**	**Total funding[b]**
Senegal Millennium Water and Sanitation Program (2009-2014)[c]	Activity focuses on governance and management, the creation of local business opportunities, increasing demand for clean water and sanitation, infrastructure construction, and community-led total sanitation for hygiene.	BCC, CB, IC, PG	Rural, peri-urban	**$20,866,000**
Yaajeende Agriculture (2010-2015)	Activity focuses on nutrition-based agriculture to improve nutritional status and includes some WASH activities to promote clean drinking water.	BCC, CB	Rural	**3,250,000**
Development Grant Program III (2012-2015)	Activity focuses on community-led total sanitation and also promotes clean drinking water.	BCC	Rural	**2,000,000**

Activity (duration)	Description	Activity focus	Geographic focus[a]	Total funding[b]
Community Health Program Component (2011-2016)	Activity supports access to a range of health services, including WASH. WASH component supports improved hygiene practices.	BCC	Rural	**800,000**
Health Communication and Promotion Program Component (2012-2016)[c]	Activity supports communications related to a range of health areas, including WASH. WASH component provided support for water purification tablets.	BCC	Rural, peri-urban, urban	**700,000**
Total activities 5				**Total funding $27,616,000**

Legend:
BCC = behavior-change communication
CB = capacity building
F = financing IC = infrastructure construction PG = policy and governance
TA = technical assistance USAID = U.S. Agency for International Development
WASH = water supply, sanitation, and hygiene
Source: GAO analysis of USAID documents and data. | GAO-16-81

[a] According to a USAID official, USAID missions do not employ a universal definition for rural, peri- urban, or urban, and missions are encouraged to defer to host-country census authorities' definitions of these terms.

[b] Funding shown generally represents allocations for WASH components of activities and includes, when applicable, WASH funding for years before fiscal year 2012 and through fiscal year 2014. We included obligations for activities for which USAID data do not show WASH allocations, such as activities for which a USAID mission had obligated unplanned funding for WASH.

[c] We selected this activity for our assessment of USAID's planning, monitoring, and—where applicable—evaluation of WASH.

Tanzania

The USAID mission in Tanzania implemented three WASH activities in fiscal years 2012 through 2014, with WASH activity funding totaling $17,753,586. The mission's primary WASH activity, called Tanzania Integrated Water, Sanitation and Hygiene Program, included $17,443,586 in funding for WASH. The activity supported community piped and gravity-fed water schemes, rehabilitated wells with rope pumps, training for community groups responsible for operations and maintenance of water schemes, and school latrines. Activity also included a water resource management component.

Table 10. USAID WASH Activities Implemented in Tanzania, Fiscal Years 2012-2014

Activity (duration)	Description	Activity focus	Geographic focus[a]	Total funding[b]
Tanzania Integrated Water, Sanitation and Hygiene Program (2010-2015)[c]	Activity supports community piped and gravity-fed water schemes, rehabilitating wells, training for community groups responsible for operations and maintenance of water schemes, and school latrines. Activities also include a water resource management component.	BCC, CB, F, IC, PG, TA	Rural, peri-urban	**$17,443,586**
Scaling Up Conservation in Maasai Steppes (2010-2013)	Activity supported investments in land, water, and resource management strategies, and included efforts related to potable water and sanitation.	BCC, PG	Rural	**300,000**
Peace Corps Volunteers' Small Projects Activities Program (2009-2012)	Activity supported institutional capacity building, investments in basic rural water infrastructure, and efforts to foster improvements in sanitation and hygiene behavior.	BCC, CB	Rural, peri-urban	**10,000**
Total activities 3				**Total funding $17,753,586**

Legend:
BCC = behavior-change communication
CB = capacity building
F = financing
IC = infrastructure construction PG = policy and governance TA = technical assistance
USAID = U.S. Agency for International Development
WASH = water supply, sanitation, and hygiene
Source: GAO analysis of USAID documents and data. | GAO-16-81

[a] According to a USAID official, USAID missions do not employ a universal definition for rural, peri-urban, or urban, and missions are encouraged to defer to host-country census authorities' definitions of these terms.

[b] Funding shown generally represents allocations for WASH components of activities and includes, when applicable, WASH funding for years before fiscal year 2012 and through fiscal year 2014. We included obligations for activities for which USAID data do not show allocations, such as activities for which a USAID mission had obligated unplanned funding for WASH.

[c] We selected this activity for our assessment of USAID's planning, monitoring, and—where applicable—evaluation of WASH.

Figure 10 shows a demonstration rope pump, which the Tanzania Integrated Water, Sanitation and Hygiene Program supported as a cost-effective, easy-to-maintain technology, and a piped water point in the village of Mvumi, Wami-Ruvu River Basin, Tanzania.

Source: GAO. | GAO-16-81

Figure 10. Tanzania's Integrated Water, Sanitation, and Hygiene Program Supported Efforts in Mvumi Village, Wami-Ruvu River Basin.

Uganda

The USAID mission in Uganda implemented 15 WASH activities in fiscal years 2012 through 2014, with WASH activity funding totaling $15,284,415. The mission's largest WASH activity, called Northern Uganda Development of Enhanced Local Governance, Infrastructure, and Livelihoods, included $6,225,000 in funding to support local government efforts in northern Uganda to expand basic WASH services. Another activity, called WASHPlus: Supportive Environments for Healthy Communities, included $500,000 in funding for WASH and aimed to build the capacity of district government and USAID implementing partners for WASH efforts to support community-led total sanitation, promoting handwashing in villages, and integrating WASH with nutrition and HIV/AIDS services and programs.

Table 11. USAID WASH Activities Implemented in Uganda, Fiscal Years 2012-2014

Activity (duration)	Description	Activity focus	Geographic focus[a]	Total funding[b]
Northern Uganda Development of Enhanced Local Governance, Infrastructure, and Livelihoods (2009-2014)	Activity supported local government efforts in northern Uganda to expand basic WASH services.	CB, IC	Rural, peri-urban, urban	**$6,225,000**
Increase Access to Potable Water in Rural Northern Uganda (2014-2016)	Activity aims to promote the development and implementation of sanitation plans in selected towns in northern Uganda.	BCC, CB, IC	Peri-urban, urban	**2,000,000**
Strengthening Decentralization for Sustainability (2010-2016)	Activity engages district governments to design, implement, and monitor WASH activities. It aims to support an enabling environment for policy reforms for improved sanitation and hygiene in districts.	BCC, CB	Rural, peri-urban, urban	**1,500,000**
Rehabilitation and Expansion of Kitgum and Pader Supply Systems (2012-2014)	Activity aimed to improve the delivery of water services in town councils. USAID supported water infrastructure construction, and the government of Uganda assumed control of the water supply after completion.	BCC, CB, IC	Peri-urban, urban	**1,000,000**
Water Infrastructure Program in Kitgum and Pader (2011-2013)	Activity focused on water infrastructure projects and included support for hardware, computer software, and training related to operations and maintenance of water supply.	CB, F, IC, PG, TA	Rural, peri-urban,urban	**1,000,000**
Northern Uganda Health Integration to Enhance Services (2012-2015)	Activity includes WASH as a component of a broader health behavior-change communication effort.	BCC	Rural, peri-urban	**550,000**

Table 11. (Continued)

Activity (duration)	Description	Activity focus	Geographic focus[a]	Total funding[b]
Community Connector (2011-2016)	Activity supports sustainable food and livelihood security through community-level action. WASH component includes support for latrines and point-of-use clean water interventions.	BCC	Rural	**500,000**
WASHPlus: Supportive Environments for Healthy Communities (2013-2014)[c]	Activity worked to build capacity of district government and USAID implementing partners for WASH, including support for community-led total sanitation, promoting hand washing in villages, and integrating WASH with nutrition and HIV/AIDS services.	CB, TA	Rural, peri-urban	**500,000**
Health Marketing Initiative (AFFORD II) (2010-2014)[c]	Activity used communication and marketing techniques to improve attitudes and knowledge about childhood diarrhea and clean water. Promoted products to treat diarrhea-related illness and purify water.	BCC	Peri-urban, urban	**450,000**
Strengthening Health Outcomes through the Private Sector (2009-2015)	Activity supports mass media campaigns related to WASH behavioral change.	BCC	Rural, peri-urban, urban	**450,000**
Water Infrastructure Consultancy and Supervision (2011-2013)	Activity provided technical assistance for water infrastructure.	CB, F, IC, PG, TA	Rural, peri-urban, urban	**409,415**
Communication for Better Health/Communication for Healthy Communities (2013-2018)	Activity designs and executes health communication interventions to improve knowledge, attitudes, behaviors, and demand for services related to WASH.	BCC	Rural, peri-urban, urban	**350,000**

Activity (duration)	Description	Activity focus	Geographic focus[a]	Total funding[b]
Strides for Family Health (2009-2015)	Activity aims to address diarrhea prevention at the facility and community levels. Interventions include training care givers on good practices.	BCC	Rural, peri-urban	**250,000**
PATH–Advocacy for Better Health (2014-2019)	Activity aims to build the capacityof communities to identify needs and to demand quality WASH services; build the capacity of organizations to advocate and represent community demands; and increase organizations' influence on decision-makers for WASH.	BCC	Rural, peri-urban, urban	**50,000**
Peace Corps Small Project Assistance Participating Agency Program Agreement (2014-2017)	Activity supports Peace Corps Volunteers' WASH activities in rural Ugandan communities.	BCC	Rural	**50,000**
Total activities 15				**Total funding $15,284,415**

Legend:

BCC = behavior-change communication

CB = capacity building

F = financing

IC = infrastructure construction PG = policy and governance TA = technical assistance

USAID = U.S. Agency for International Development

WASH = water supply, sanitation, and hygiene

Source: GAO analysis of USAID documents and data. | GAO-16-81

[a] According to a USAID official, USAID missions do not employ a universal definition for rural, peri- urban, or urban, and missions are encouraged to defer to host-country census authorities' definitions of these terms.

[b] Funding shown generally represents allocations for WASH components of activities and includes, when applicable, WASH funding for years before fiscal year 2012 and through fiscal year 2014. We included obligations for activities for which USAID data do not show WASH allocations, such as activities for which a USAID mission had obligated unplanned funding for WASH.

[c] We selected this activity for our assessment of USAID's planning, monitoring, and—where applicable—evaluation of WASH. We did not select the Uganda mission's largest WASH activity, Northern Uganda Development of Enhanced Local Governance, Infrastructure, and Livelihoods, for our assessment, because the activity received the majority of its funding before fiscal year 2012, and the mission did not initially inform us of the activity and did not provide prior year funding data for the activity until after we had made our activity selection. We based our activity selection on funding data reported in State's Foreign Assistance Coordination and Tracking System for fiscal years 2012 and 2013, on data included in missions' operational plans for fiscal year 2014, and on discussions with mission officials about their ongoing WASH activities in fiscal years 2012 through 2014.

Zambia

The USAID mission in Zambia implemented three WASH activities in fiscal years 2012 through 2014, with WASH activity funding totaling $27,600,000. The mission's largest WASH activity, called Schools Promoting Learning Achievement through Sanitation and Hygiene, included $15,200,000 in funding for WASH and focused on providing WASH services in schools. Efforts included hygiene education, capacity building for operations and maintenance, and support to establish private- sector spare-parts supply.

Table 12. USAID WASH Activities Implemented in Zambia, Fiscal Years 2012-2014

Activity (duration)	Description	Activity focus	Geographic focus[a]	Total funding[b]
Schools Promoting Learning Achievement through Sanitation and Hygiene (2011-2015)[c]	Activity provides WASH services in schools, hygiene education, capacity building for operations and maintenance, and support for private sector spare- parts supply.	BCC, CB, IC, PG, TA	Rural, peri-urban, urban	**$15,200,000**
Partnership for Integrated Social Marketing (2009-2014)[c]	Activity distributed point-of-use water disinfecting solution.	BCC, CB, F, PG, TA	Rural, peri-urban, urban	**6,900,000**
School WASH (2009-2012)	Activity provided WASH services in schools, hygiene awareness training, and capacity building for school WASH committees related to operations and maintenance.	BCC, CB, IC, PG, TA	Rural, peri-urban, urban	**5,500,000**
Total activities 3				**Total funding $27,600,000**

Legend:
BCC = behavior-change communication
CB = capacity building
F = financing
IC = infrastructure construction PG = policy and governance TA = technical assistance
USAID = U.S. Agency for International Development
WASH = water supply, sanitation, and hygiene
Source: GAO analysis of USAID documents and data. | GAO-16-81
[a] According to a USAID official, USAID missions do not employ a universal definition for rural, peri- urban, or urban, and missions are encouraged to defer to host-country census authorities' definitions of these terms.

[b] Funding shown generally represents allocations for WASH components of activities and includes, when applicable, WASH funding for years before fiscal year 2012 and through fiscal year 2014. We included obligations for activities for which USAID data do not show WASH allocations, such as activities for which a USAID mission obligated unplanned funding for WASH.

[c] We selected this activity for our assessment of USAID's planning, monitoring, and—where applicable—evaluation of WASH.

APPENDIX III: USAID EVALUATIONS' ASSESSMENT OF MONITORING, OUTCOMES, AND SUSTAINABILITY

Our review of 14 U.S. Agency for International Development (USAID) evaluations of water, sanitation, and hygiene (WASH) activities in nine selected countries found that the evaluations assessed monitoring, outcomes, and sustainability to varying extents.[68]

Monitoring. Evaluations reported on various issues, including monitoring, WASH indicators, indicators' limitations, and data related to gender.

- All 14 evaluations reported on monitoring of WASH activities.
- Ten evaluations reported on one or more of USAID's indicators related to access to drinking water, sanitation, and hygiene, including whether WASH activities were on track to meet their targets.
- Five evaluations discussed limitations in the quality of one or more indicators used to monitor WASH activities. For example, an evaluation of Senegal's Yaajeende activity noted limitations of the indicator for the number of individuals trained on improved hygiene behaviors. The evaluation stated that, according to USAID's guidance, the success of training and other interventions related to human and organizational capacity building is to be measured by improvement in organizational output and performance, not simply by the number of individuals trained.
- Five evaluations presented and discussed data related to gender, thereby filling some gaps that we had identified related to a lack of disaggregated gender data in monitoring reports. For example, the evaluations for the primary WASH activities in Ethiopia and Senegal assessed women's participation in community water management committees, which were generally responsible for operations and maintenance (see Figure 11).[69]

Outcomes. Six of the 14 evaluations that we reviewed provided insights into WASH activity outcomes. Specifically, these evaluations had findings related to outcomes such as disease incidence, health expenses, school attendance, time spent getting water, economic impacts, and beneficiary satisfaction. For example, the evaluation of Ethiopia's Water Sanitation and Hygiene Transformation for Enhanced Resiliency found that the activity resulted in access to safe water at a much closer distance than before and also increased access to safe latrines and improved health practices (such as handwashing). In addition, the evaluation found that activity results included increased time for beneficiaries to participate in other productive and income-generating activities, including more time at school, as well as reduced health expenses.

Source: GAO. | GAO-16-81

Figure 11. USAID-Funded Water Point in Ethiopia with One of Three Female Members of a Seven-Person Water Management Committee.

Sustainability. Twelve of the 14 evaluations that we reviewed addressed the sustainability of WASH activities. Specifically, these evaluations broadly discussed WASH sustainability issues, and 7 of the 12 had findings related to WASH sustainability challenges. These challenges included limitations related to capacity building, a lack of spare parts, and a lack of funding for operations and maintenance. For example, the evaluation of Tanzania's Integrated Water, Sanitation, and Hygiene Program reported on factors that improved sustainability, such as the ease of repairing a rope pump. The Tanzania evaluation also noted challenges related to capacity building for community water committees. The evaluation assessed 10 of 26 committees as having "fair" usage and maintenance, where community fees were generally not collected or maintenance was spotty, and rated 6 of the 26 as "poor" for underperformance relative to the rest of the project.

End Notes

[1] Centers for Disease Control and Prevention, *Global WASH Fast Facts: Information on Water, Sanitation, and Hygiene* (Atlanta, Georgia: June 2015), accessed Aug. 18, 2015, http://www.cdc.gov/healthywater/global/wash_statistics.html

[2] Pub. L. No. 109-121.

[3] In September 2010, we reported that State needed to strengthen its strategic approach to U.S. water and sanitation assistance and recommended that State, in consultation with USAID, (1) ensure that the U.S. water and sanitation strategy addressed all components required by the 2005 Act and (2) explain the basis for its designations of priority countries for U.S. assistance. State concurred with our recommendations. In July 2014, State's Special Coordinator for Water informed us that State had not developed a U.S. government-wide water and sanitation strategy but, because USAID administers the majority of U.S. funding for water and sanitation assistance, had supported the development of USAID's *Water and Development Strategy, 2013-2018* (Water Strategy), which was released in May 2013. Because the 2005 Act called for State to develop a U.S.-government-wide strategy, we were unable to close the first recommendation as implemented. In July 2014, we closed the second recommendation as implemented, because we found that USAID's Water Strategy includes clearly articulated criteria for selecting priority countries. GAO, *U.S. Water and Sanitation Aid: Millions of Beneficiaries in Developing Countries, but Department of State Needs to Strengthen Strategic Approach*, GAO-10-957 (Washington, D.C.: Sept. 24, 2010).

[4] Pub. L. No. 113-289.

[5] We reported that USAID, the Millennium Challenge Corporation, and 23 other agencies funded international water and sanitation assistance, with USAID providing the majority of WASH funding. See *GAO, Foreign Assistance: Briefing on U.S. International Water- Related Assistance,* GAO-14-683R (Washington, D.C.: July 24, 2014). The United States also provides significant contributions to international financial institutions and international organizations, which support international water and sanitation projects. Assistance provided through multilateral institutions was not within the scope of this review.

[6] In fiscal year 2014, USAID designated 22 countries as priority for WASH assistance based on the level of need in these countries and opportunities for USAID projects to have an impact: Ethiopia, Kenya, Indonesia, Liberia, Nigeria, and South Sudan (tier 1); and Bangladesh, Cambodia, DRC, Ghana, Haiti, India, Malawi, Mali, Mozambique, Nepal, the Philippines, Rwanda, Senegal, Tanzania, Uganda, and Zambia (tier 2). According to USAID, tier 1 countries have the potential to significantly leverage USAID resources to implement WASH projects at scale. USAID anticipates that tier 2 countries will require relatively small investment levels to achieve a significant impact in at least one dimension of WASH.

[7] Annual appropriations acts funding foreign operations have established spending requirements for WASH since the 2005 Act was passed. Funding data for fiscal year 2014 were not available at the time that we made our country selection.

[8] We reviewed USAID guidance including: (1) USAID's Automated Directives System (ADS), ch. 201, which contains the agency's policies and procedures and includes guidance on strategic planning and project and activity design; (2) USAID's Water Strategy and (3) USAID's *Water and Development Strategy: Implementation Field Guide* (March 2014).

[9] ADS ch. 201 defines an activity as a component of a project that contributes to a project purpose; a project is a set of executed interventions over an established timeline and with a budget that are intended to achieve a discrete development result. The selected 16 activities

generally represented one award each (e.g., cooperative agreement, grant, or contract), with one exception (the Haiti activity represents two contracts).

[10] Tanzania had only one primary WASH activity. We initially identified three WASH activities in Haiti; however, USAID's mission in Haiti subsequently informed us that two of the selected activities did not have a WASH component and that the mission had incorrectly attributed WASH funds to these activities. The mission planned to take corrective action to ensure use of allocated funds for WASH. As a result, we reviewed one WASH activity for Haiti.

[11] According to USAID, the spending requirement was reduced to $302 million in fiscal year 2013 because of sequestration. Funding attributed to the annual spending requirements for WASH must be from accounts funded by Foreign Operations appropriations measures and must adhere to the various statutory requirements imposed on those accounts by the Foreign Assistance Act of 1961 (Pub. L. No. 87–195, as amended).

[12] This annual guidance is called State's Water Key Issue Definition.

[13] The Water Strategy included a second strategic objective, to manage water in agriculture sustainably and more productively to enhance food security. We did not include the Water Strategy's second strategic objective within the scope of our review.

[14] For the purposes of this review, we refer to the Water Strategy's intermediate results as goals.

[15] An improved drinking-water source is one that, by the nature of its construction and when properly used, adequately protects the source from outside contamination, particularly fecal matter.

[16] An improved sanitation facility is one that hygienically separates human excreta from human contact.

[17] The Water Strategy puts an emphasis on scaling up and strengthening sanitation programs. For example, the strategy highlights approaches such as community-led total sanitation (CLTS), which focuses on changing collective behavior in communities, rather than on building donor-funded latrines for each household, to end open defecation. Additionally, the Water Strategy notes that USAID missions can promote community or public toilets in dense settlements as well as sanitation facilities in institutional settings such as schools, health facilities, and transportation hubs.

[18] These indicators are consistent with the United Nations' (UN) two primary goals for WASH. Specifically, the UN adopted Millennium Development Goals to halve, by 2015, the proportion of the world's population that lacks sustainable access to safe drinking water and basic sanitation. The UN World Health Organization (WHO) and the UN Children's Fund (UNICEF) are responsible for monitoring progress toward this goal.

[19] State's fiscal year 2013 annual report to Congress on WASH notes that 3.5 million people gained access to an improved drinking water source and 1.29 million people gained access to an improved sanitation facility. Office of Conservation and Water, Bureau of Oceans and International Environmental and Scientific Affairs, U.S. Department of State, *Annual Report to Congress: Senator Paul Simon Water for the Poor Act P.L. 109-121; Sec. 6(g)(2)* (Washington, D.C.: June 2014).

[20] The Field Guide notes that missions should also consider associating WASH activities with improvements in other health objectives, such as nutrition. Further, it states that the Water Strategy's second strategic objective addresses critical linkages between USAID's objectives for the water sector and for its Feed the Future program, which include inclusive agricultural growth, nutrition, and climate change. We previously reported on USAID's global food security efforts; see GAO, *Global Food Security: USAID Is Improving Coordination but Needs to Require Systematic Assessments of Country-Level Risks*, GAO-13-809 (Washington, D.C.: Sept. 17, 2013).

[21] Water-related services include those funded by user fees such as water and wastewater user fees. USAID Policy, *Sustainable Service Delivery in an Increasingly Urbanized World* (Washington, D.C.: October 2013.) According to USAID, the agency considers governance,

financial, technical, social, and environmental factors in designing, planning, implementing, and monitoring WASH projects for sustainability.

[22] A third priority tier consists of 6 countries or locations where USAID anticipates continued WASH projects because of a combination of strategic considerations and development needs: Afghanistan, Jordan, Lebanon, Pakistan, West Bank/Gaza, and Yemen. In these countries, strategic priorities other than low levels of access to improved drinking water and basic sanitation or associated health impacts drive USAID funding of WASH activities. USAID reported attributing about $369.8 million to the spending requirements for WASH in the 6 tier 3 priority countries in fiscal years 2012 through 2013.

[23] We identified these focus areas based on review of USAID documents and information and discussions with USAID officials.

[24] Some activities started before fiscal year 2012, and some were ongoing at the end of fiscal year 2014. Funding shown generally represents total allocations for WASH activities ongoing at each mission in fiscal years 2012 through 2014, including, when applicable, funding for years before fiscal year 2012 and through fiscal year 2014. We included obligations in cases where USAID data do not show WASH activity allocations, such as when a USAID mission obligated previously unplanned funding to an activity for WASH; we included total obligations of about $4.3 million for 6 WASH activities implemented by 4 missions. State and USAID define allocations as the distribution of resources to bureaus and operating units by foreign assistance account. For the purposes of this review, we defined obligations as orders placed, contracts awarded, and similar transactions during a given period that will require payments during the same or a future period. USAID categorizes these as "sub-obligations", because it considers these funds to have been obligated through a bilateral agreement with the host country.

[25] In some cases, activities focused primarily on WASH, while in other cases, activities included WASH in addition to other components, such as maternal-child health, nutrition, or natural resources management.

[26] Missions reported multiple focus areas for most WASH activities.

[27] According to a USAID official, USAID missions do not employ a universal definition for rural, peri-urban, and urban, and missions are encouraged to defer to national census authorities' definitions of these terms.

[28] For the purposes of this report, we define WASH plans principally as documentation describing a mission's WASH project design, which may be contained in one or more PADs related to WASH, as well as mission-level WASH sector assessments or other related documents that describe the mission's strategic approach to WASH. ADS ch. 201 (effective July 2, 2013) requires missions to develop a PAD documenting a project's design and to align activities and expected results with the mission's broader results framework as part of the mission's project design process. Components of the PAD include the development problem, technical approach, expected results, and analytical and sustainability considerations, among other components. USAID officials noted that the PAD is broader and may encompass multiple interventions, whereas the activity approval document that USAID previously required was activity specific. We found that, because WASH is a cross-cutting issue, a PAD may focus on WASH or include WASH as a component of a broader project, such as education.

[29] These missions completed their latest WASH plans in fiscal years 2015 and 2014, respectively.

[30] We previously reported that USAID had not provided clear guidance to missions for completing what is known as a 611(e) certification, as required by the Foreign Assistance Act of 1961, attesting to a host country's capability to effectively maintain and utilize any capital assistance project whose estimated costs exceed $1 million, and that partly as a result, USAID missions have failed to complete the required certifications for many projects. Our report noted that USAID had set up working groups to discuss the 611(e) certification requirement, and we recommended that USAID provide guidance for identifying activities for which sustainability must be certified and specifying information

that these certifications should include. GAO, *Haiti Reconstruction: USAID Has Achieved Mixed Results and Should Enhance Sustainability Planning,* GAO-15-517 (Washington, D.C.: June 3, 2015). We did not assess the completion of 611(e) certification requirements for selected missions' WASH activities as part of this review.

[31] Missions are generally required to include a sustainability analysis related to WASH as part of a PAD. ADS (ch. 201) requires missions to assess sustainability as part of the process of developing a PAD, which includes analysis of sustainability factors related to issues including economy, finance, gender, culture, and institutional capacity, as well as technical and environmental considerations.

[32] The award agreement for a WASH activity in Tanzania included a section describing the activity's sustainability and exit plan, which noted that long-term sustainability would be the "cornerstone" of the activity. Among other things, the agreement stated that implementing partners would take steps to help ensure sustainability, including empowering local communities to decide on their water needs and services to meet those needs, requiring communities to contribute toward financing and construction of water systems, establishing governance structures to manage and maintain water systems, and involving the private sector to develop livelihoods and a local water services supply chain.

[33] As another example, the Uganda mission also completed a sustainability assessment in September 2014 for one of its ongoing WASH activities to identify ways to address challenges and enhance efforts related to community mobilization and operations and maintenance of WASH facilities and infrastructure.

[34] According to USAID officials, USAID has applied the sustainability assessment tool in a fourth country, Ethiopia, however, results were not provided to GAO as of September 2015.

[35] For example, the sustainability assessment for Kenya identified sustainability challenges for some activities, including insufficient community funding to support operations and maintenance, highly variable spare parts supply, and low gender balance on some water committees. In addition, almost half of households in target communities traveled more than 1 kilometer or waited more than an hour for water, a fact that could affect beneficiaries' willingness to maintain projects over the long term. Among other things, the assessment included recommendations for the mission to work with national and district stakeholders to meet WASH needs; support the role of women on water committees; and consult with communities early on regarding placement, applicability, and management of services to be provided.

[36] In addition, USAID officials noted that the agency is in the process of revising agencywide guidance related to sustainability. The officials also noted that USAID intends to make sustainability a central focus of project design, consistent with the approach to sustainability outlined in USAID's Local Systems Framework, which calls for designing, monitoring, and evaluating interventions for sustainable results. For more information, see USAID, *Local Systems: A Framework for Supporting Sustained Development* (Washington, D.C.: April 2014.)

[37] In 2015, UNICEF published a concept note regarding the need to strengthen WASH governance through accountability to achieve sustainable WASH. The concept note stated that WASH interventions frequently fail to bring sustainable benefits, posing a major obstacle to universal access. It added that UNICEF will develop guidance on how accountability can be reinforced through WASH programming. UNICEF, *Accountability in WASH: Explaining the Concept* (Stockholm, Sweden: 2015). WHO and UNICEF are in the process of determining post-2015 Sustainable Development Goals focusing on universal and equitable access for WASH, to replace the 2015 Millennium Development Goals.

[38] According to USAID policy (ADS ch. 203), performance monitoring and evaluation are mutually reinforcing, but distinct, practices. Performance monitoring is an ongoing process that indicates whether desired results are occurring and whether the development objective and project outcomes are on track. It uses preselected indicators to measure progress toward planned results on a continuous basis. Evaluation is the systematic collection and analysis of

information about the characteristics and outcomes of programs and projects as a basis for judgments to improve effectiveness, and inform decisions about current and future programming. It provides an opportunity to consider both planned and unplanned results and to reexamine the development hypothesis of the development objective and to make adjustments based on new evidence. USAID mission personnel have responsibility for oversight of the implementing organization's activity-level monitoring. USAID requires that most evaluations be conducted by external experts.

[39] In this report, "key WASH indicators" refers to WASH indicators whose use is recommended in the Field Guide. These indicators include number of people gaining access to an improved drinking water source, number of people gaining access to an improved sanitation facility, and number of liters of drinking water disinfected with point-of- use treatment products. If an activity did not use WASH indicators recommended in the Field Guide, we selected at least one of the activity's indicators related to drinking water, sanitation, and hygiene efforts, as applicable. A complete suite of WASH indicators can be found in the Field Guide.

[40] USAID ADS 203 (revised in Nov. 2012). A previous version of ADS (revised in April 2010) also required that for each indicator in a performance management plan, a USAID mission should include performance baselines and set performance targets. Additionally, this version of ADS noted that annual targets are generally required for the standard indicators selected for the annual operational plans and the performance reports from the missions.

[41] The annual reports for 13 of the 16 activities we reviewed included performance data for key WASH indicators.

[42] USAID ADS 203 (revised in Nov. 2012) requires that all people-level indicators at the mission strategy, project, or activity level be sex disaggregated to help USAID assess contributions to gender equality and female empowerment. Previous version of ADS 203 (revised April 2010) also required that USAID missions include gender-sensitive indicators and sex-disaggregated data when the analyses supporting an objective, project, or activity demonstrated that the different roles and status of women and men within the community, political sphere, workplace, and household affected the activities to be undertaken. This report uses the terms gender and sex interchangeably.

[43] Of the 16 activities that we reviewed, 10 included a water component and 8 included a sanitation component.

[44] Consistent with WHO and UNICEF definitions, USAID considers a shared water source, such as a public tap or tube well, an improved drinking water source. State and USAID guidance allows USAID missions to estimate beneficiaries accessing an improved drinking water source but requires that on completion of construction or rehabilitation of an improved water source, the USAID implementer or third-party evaluator estimate beneficiaries by verifying usage through observations and interviews with initial users and by taking into account time to collect water and quantity of water produced by the water source. According to State and USAID guidance for the indicator of the number of people accessing an improved drinking water source, the beneficiaries should spend 30 minutes or less to collect water from the improved drinking water source. Additionally, the water source should generate enough water to allow each beneficiary to consume at least 20 liters of water per day. The guidance recognizes that this indicator can be difficult and time consuming to measure accurately; therefore, it requires robust data quality assurance on the part of USAID.

[45] The Tanzania mission's estimate of beneficiaries of its largest WASH activity was initially based on the assumption that one water pump benefitted 250 people. The mission subsequently undertook a verification study that indicated that the number of beneficiaries was overestimated and that the actual number of users per pump was closer to 150. On the basis of the study's results, the mission lowered its estimate of the activity's beneficiaries. For its current primary WASH activity, the Ethiopia mission commissioned a midterm evaluation that confirmed the planning estimate of about 20 liters of water per beneficiary

per day. For its current primary WASH activity, officials at the Indonesia mission stated that they undertook verification of data for indicators including access to improved drinking water sources. Additionally, the midterm evaluation for this activity noted that results were reasonably attributed and can be verified.

[46] The Senegal mission reported in fiscal years 2012 and 2013 that 17,190 and 19,860 people, respectively, gained access to an improved drinking water source as a result of the mission's primary WASH activity. Mission officials stated that the calculation of these estimates conformed with host-government guidance that one mini-borehole or rehabilitated well provides access to 150 people, while one medium-sized or deep mini- borehole provides access to 300 people.

[47] The annual reports for the DRC mission's primary WASH activity reported that 793,922 people gained access to an improved drinking water source between fiscal years 2012 and 2014. The mission's second WASH activity we reviewed reported 123,464 beneficiaries gaining access to an improved drinking water source as of October 2014.

[48] In its Water Strategy, USAID noted that it would increase its focus on sanitation given more limited progress worldwide on access to improved sanitation compared with access to improved drinking water sources. USAID's primary indicator to assess progress on sanitation is the number of people gaining access to an improved sanitation facility.

[49] According to USAID, basic latrines (such as a pit latrine without a slab) can count toward the indicator for reducing open defecation. USAID also has a separate indicator to measure the number of improved toilets in institutional settings.

[50] For its primary WASH activity, the Senegal mission used separate sanitation indicators to report progress on (1) construction of basic latrines under the community-led total sanitation effort (indicator reported: number of villages attaining open defecation free status in 6 months), (2) construction of shared sanitation facilities (indicator reported: number of improved toilets provided in institutional settings), and (3) construction of improved sanitation facilities for households (indicator reported: number of people gaining access to an improved sanitation facility).

[51] The monitoring reports for this DRC mission activity indicate that 43,980 (in fiscal year 2012), 249,678 (in fiscal year 2013), and 228,543 (in fiscal year 2014) beneficiaries gained access to an improved sanitation facility. In USAID's annual report for fiscal year 2013, the number of beneficiaries reported as gaining access to an improved sanitation facility in the DRC constituted 25 percent of worldwide beneficiaries (320,924 out of 1,299,023).

[52] We did not assess the extent to which latrines built under this activity included slabs.

[53] In USAID's annual report for fiscal year 2013, the number of beneficiaries reported as gaining access to an improved sanitation facility in Zambia constituted 9 percent of worldwide beneficiaries (110,737 of 1,299,023).

[54] According to the Zambia mission, all schools where toilets were built under this activity have two or three sessions for students. For example, students in some grades might attend schools in the morning, while students in other grades might attend school in the afternoon. The mission calculated the ratio of students to toilets based on the session with the largest student population rather than the total school population because not all students attend school at the same time.

[55] USAID requires that each USAID mission or office commission at least one evaluation of each large project and of each pilot or innovative project it implements. Furthermore, USAID requires that these external entities, other than USAID and activity implementer, conduct these evaluations (ADS, ch. 203). We did not assess selected USAID missions' compliance with the requirement to conduct evaluations.

[56] These 14 evaluations represent all evaluations that USAID completed for WASH activities in the nine selected countries in fiscal years 2012 through 2014.

[57] These factors are consistent with USAID guidance, which supplements USAID's policy on evaluation. This guidance consists of a checklist of 40 factors, including the following topics: adherence to general principles in USAID's evaluation policy; identification of the

activity, project, or approach to be evaluated; identification of existing performance information source, with special attention to monitoring data; stating the purpose of, audience for, and use of the evaluation; identification of the evaluation methods; specification of evaluation deliverables and the timeline; and discussion of evaluation team composition (one team member should be an evaluation specialist) and participation of customers and partner.

[58] In one case, an activity in our review was modified based on an evaluation of an earlier activity that was not part of our review.

[59] USAID plans to initiate an annual series of long-term evaluations of projects in different sectors, each year examining a different set of projects with similar aims to understand their lasting effects. USAID, *Local Systems: A Framework for Supporting Sustained Development* (Washington, D.C.: April 2014).

[60] This is the second of two reports in response to a congressional request that we review the U.S. government's international water-related assistance. In our first report, issued in July 2014, we provided information regarding U.S. agencies' funding, roles and responsibilities, staffing, and coordination and collaboration on international water-related assistance. *GAO, Foreign Assistance: Briefing on U.S. International Water-Related Assistance,* GAO-14-683R (Washington, D.C.: July 24, 2014).

[61] In July 2014, we reported that USAID had provided WASH assistance in a total of 85 countries in fiscal years 2009 through 2013. See GAO-14-683R.

[62] In fiscal year 2014, USAID designated 22 countries as priorities for WASH assistance: Ethiopia, Kenya, Indonesia, Liberia, Nigeria, and South Sudan (tier 1); and Bangladesh, Cambodia, the DRC, Ghana, Haiti, India, Malawi, Mali, Mozambique, Nepal, the Philippines, Rwanda, Senegal, Tanzania, Uganda, and Zambia (tier 2). According to USAID, tier 1 countries have the potential to significantly leverage USAID resources to implement WASH projects at scale. USAID anticipates that tier 2 countries will require relatively small investment levels to achieve significant impact in at least one dimension of WASH.

[63] Funding data for fiscal year 2014 were not available when we made our country selection.

[64] USAID reported attributing about $743 million in total funding to the congressional spending requirement for international water and sanitation assistance in fiscal years 2012 through 2013. The 9 selected countries accounted for about 21 percent of this funding.

[65] State and USAID define allocations as the distribution of resources to bureaus and operating units by foreign assistance account. For the purposes of this review, we defined obligations as orders placed, contracts awarded, and similar transactions during a given period that will require payments during the same or a future period. USAID categorizes these as sub-obligations, because it considers these funds to have been obligated through a bilateral agreement with the host country.

[66] To assess if WASH activity funding was appropriately attributed, we compared information we obtained from missions on selected activities to State's annual Water Key Issue Definition guidance.

[67] USAID's Automated Directive System (ch. 203) requires that each USAID mission or office conduct at least one evaluation of each large project it implements. The policy defines a large project as one that equals or exceeds in dollar value the average project size for each development objective for that USAID mission or office. Additionally, USAID recommends evaluation of any activity within a project involving untested hypotheses or demonstrating new approaches that are anticipated to be expanded in scale or scope through U.S. government foreign assistance or other funding sources.

[68] USAID missions commissioned the 14 evaluations for, respectively, one activity in the Democratic Republic of the Congo, two activities in Ethiopia, one activity in Haiti, three activities in Indonesia, one activity in Kenya, two activities in Senegal, two activities in Tanzania, one activity in Uganda, and one activity in Zambia.

[69] The evaluation for USAID's major WASH activity in Senegal noted that the participation of women in water-users associations was noticeable and that women held decision- making positions such as that of vice president, which is reserved for them. The report noted that this position was filled in 57 percent of the water-users associations. The evaluation for USAID's major WASH activity in Ethiopia noted that approximately one-third of the members of water management committees were women, who worked as cashiers and helped to address villagers' concerns and grievances. The evaluation noted that women, who were generally responsible for fetching water, were also well represented as beneficiaries of the activity.

In: U.S. International Water ...
Editor: Constance Alvarado
ISBN: 978-1-63485-066-7

Chapter 3

GLOBAL WASH FAST FACTS*

Centers for Disease Control and Prevention

ACCESS TO WASH

- Worldwide, 780 million people do not have access to an improved water source. [# 2]
- An estimated 2.5 billion people lack access to improved sanitation (more than 35% of the world's population).[1], [3]
- According to the World Health Organization and UNICEF, regions with the lowest coverage of "improved" sanitation in 2006 were sub-Saharan Africa (31%), Southern Asia (33%) and Eastern Asia (65%).[2]
- In 2006, 7 out of 10 people without access to improved sanitation were rural inhabitants.[2]
- According to the United Nations and UNICEF, one in five girls of primary-school age are not in school, compared to one in six boys. One factor accounting for this difference is the lack of sanitation facilities for girls reaching puberty. Girls are also more likely to be responsible for collecting water for their family, making it difficult for them to attend school during school hours.[13], [14] The installation of toilets and latrines may enable school children, especially

* This is an edited, reformatted and augmented version of the following CDC website (viewed February 2016): http://www.cdc.gov/healthywater/global/wash_statistics.html.

menstruating girls, to further their education by remaining in the school system.

DISEASE & DEATH

- An estimated 801,000 children younger than 5 years of age perish from diarrhea each year, mostly in developing countries. This amounts to 11% of the 7.6 million deaths of children under the age of five and means that about 2,200 children are dying every day as a result of diarrheal diseases. [4]
- Unsafe drinking water, inadequate availability of water for hygiene, and lack of access to sanitation together contribute to about 88% of deaths from diarrheal diseases.[1]
- Worldwide, millions of people are infected with neglected tropical diseases (NTDs), many of which are water and/or hygiene-related, such as Guinea Worm Disease, Buruli Ulcer, Trachoma, and Schistosomiasis. These diseases are most often found in places with unsafe drinking water, poor sanitation, and insufficient hygiene practices.[8,9]
- Worldwide, soil-transmitted helminths infect more than one billion people due to a lack of adequate sanitation.[10]
- Guinea Worm Disease (GWD) is an extremely painful parasitic infection spread through contaminated drinking water. GWD is characterized by spaghetti-like worms up to 1 meter in length slowly emerging from the human body through blisters on the skin anywhere on the body but usually on the lower legs or lower arms. Infection affects poor communities in remote parts of Africa that do not have safe water to drink. In 2014, 126 cases of Guinea Worm Disease were reported. Most of those cases were from Sudan (56%).[11]
- Trachoma is the world's leading cause of preventable blindness and results from poor hygiene and sanitation. Approximately 41 million people suffer from active trachoma and nearly 10 million people are visually impaired or irreversibly blind as a result of trachoma.[12] Trachoma infection can be prevented through increased facial cleanliness with soap and clean water, and improved sanitation.

PREVENTION

- Water, sanitation and hygiene has the potential to prevent at least 9.1% of the global disease burden and 6.3% of all deaths [1].The impact of clean water technologies on public health in the U.S. is estimated to have had a rate of return of 23 to 1 for investments in water filtration and chlorination during the first half of the 20th century.[5]
- Water and sanitation interventions are cost effective across all world regions. These interventions were demonstrated to produce economic benefits ranging from US$ 5 to US$ 46 per US$ 1 invested [6]
- Improved water sources reduce diarrhea morbidity by 21%; improved sanitation reduces diarrhea morbidity by 37.5%; and the simple act of washing hands at critical times can reduce the number of diarrhea cases by as much as 35%. Improvement of drinking-water quality, such as point-of-use disinfection, would lead to a 45% reduction of diarrhea episodes.[7]
- In order to meet the United Nations' Millennium Development Goal[13] to halve the proportion of people without sustainable access to improved drinking water and basic sanitation by 2015:
    - An estimated 784 million people will need to gain access to an improved water source.
    - An estimated 173 million people on average per year will need to begin using improved sanitation facilities (accounting for expected population growth).[2]
- Even if the United Nations' Millennium Development Goal 13 for improved drinking water and basic sanitation is reached by 2015, it will still leave:
    - An estimated 790 million people (11% of the world's population) without access to an improved water supply.
    - An estimated 1.8 billion people (25% of the world's population) without access to adequate sanitation 15.

An improved water source is defined as water that is supplied through a household connection, public standpipe, borehole well, protected dug well, protected spring, or rainwater collection.

End Notes

[1] Prüss-Üstün A., Bos, R., Gore, F. & Bartram, J. 2008. Safer water, better health: costs, benefits and sustainability of interventions to protect and promote health. World Health Organization, Geneva.

[2] World Health Organization and UNICEF. Progress on Drinking Water and Sanitation: 2012 Update. United States: WHO/UNICEF Joint Monitoring Programme for Water Supply and Sanitation; 2012.

[3] U.S. Census Bureau. International Programs Center: Population Clocks.

[4] Liu L, Johnson HL, Cousens S, Perin J, Scott S, Lawn JE, Rudan I, Campbell H, Cibulskis R, Li M, Mathers C, Black RE; Child Health Epidemiology Reference Group of WHO and UNICEF. Global, regional, and national causes of child mortality: an updated systematic analysis for 2010 with time trends since 2000. Lancet. 2012 Jun 9;379(9832):2151-61.

[5] Cutler D, Miller G. 2004. The role of public health improvements in health advances: the 20th century United States. National Bureau of Economic Research. Working Paper 10511. Cambridge, MA, USA.

[6] Hutton G, Haller L , Bartram J. 2007. Global cost-benefit analysis of water supply and sanitation interventions. J Water Health 2007:5.4;481-502.

[7] United Nations Millennium Project. Health, Dignity, and Development: What Will it Take?

[8] World Health Organization. Neglected Tropical Diseases, Hidden Successes, Emerging Opportunities.

[9] Hotez PJ, Molyneux DH, Fenwick A, Ottesen E, Ehrlich Sachs S, Sachs JD. Incorporating a rapid-impact package for neglected tropical diseases with programs for HIV/AIDS, tuberculosis, malaria. PLoS Med 2006;3(5):e102.

[10] World Health Organization. Soil-Transmitted Helminths.

[11] WHO Collaborating Center for Research Training and Eradication of Dracunculiasis, Guinea Worm Wrap Up #232, 2015, Centers for Disease Control and Prevention (CGH): Atlanta.

[12] International Trachoma Initiative. What is Trachoma?

[13] The United Nations. Millennium Development Goals Report 2007.

[14] UNICEF and IRC. Water Sanitation and Hygiene Education for Schools: Roundtable Proceedings and Framework for Action.

[15] World Health Organization and UNICEF. Meeting the MDG Drinking Water and Sanitation Target: The Urban and Rural Challenge of the Decade.

INDEX

#

A

B

C

D

E

H

I

J

K

L

Q

R

S

T

U

V

W

Y